T0135641

18

BISS MONOGRAPHS
MONOGRAPHS OF THE
BREMEN INSTITUTE
OF SAFE SYSTEMS

R. Drechsler, M. Gogolla, H.-J. Kreowski,
B. Krieg-Brückner, J. Peleska (Series Editors)

THE BREMEN AUTONOMOUS WHEELCHAIR "ROLLAND": SELF-LOCALIZATION AND SHARED CONTROL

Axel Lankenau

Vom Fachbereich Mathematik und Informatik
der Universität Bremen
zur Verleihung des akademischen Grades eines
Doktors der Ingenieurwissenschaften (Dr.-Ing.)
genehmigte Dissertation

Gutachter: Prof. Dr. Bernd Krieg-Brückner
                  Prof. Christian Freksa (Ph.D.)

Kolloquium: 29. November 2002

Bibliografische Information Der Deutschen Bibliothek

Die Deutsche Bibliothek verzeichnet diese Publikation in der Deutschen
Nationalbibliografie; detaillierte bibliografische Daten sind im Internet über
http://dnb.ddb.de abrufbar.

ISBN 3-8325-0306-4

Logos Verlag Berlin
Comeniushof, Gubener Str. 47,
10243 Berlin
Tel.: +49 030 42 85 10 90
Fax: +49 030 42 85 10 92
INTERNET: http://www.logos-verlag.de

Summary

The Bremen Autonomous Wheelchair "Rolland" is a smart wheelchair robot intended to support handicapped and elderly people. Rolland and other service robots differ from classical industrial robots in many ways. The two most fundamental characteristics are: First, Rolland is mobile and has to navigate in environments often optimized for humans but not for robots. A prerequisite for mobility and navigation is being able to localize oneself in the environment. And second, Rolland is simultaneously controlled by the user *and* by the technical system with interdependently changing priorities: it is a so-called "shared-control system". As a consequence, two primary goals are to be pursued for Rolland: it needs an "adequate" self-localization approach. And, the human-robot interaction has to be realized in a way such that conflict situations are avoided.

Having emerged in the 1980s, the self-localization of mobile robots is still a hot research topic. It can be considered to be solved for small- and middle-scale environments provided that adequate sensor equipment (e. g. laser range finders) is available. But there are still a lot of open problems in mobile robot self-localization. Some of them are especially relevant for the service robotics domain: low-cost solutions (a $5,000 laser range finder almost doubles the price of a wheelchair) are to be found that work in large-scale environments (e. g. hospital or university buildings spread over a campus area).

This thesis presents RouteLoc, a new self-localization approach that needs only minimal input (i. e. no expensive proximity sensors are required) to absolutely localize a robot even in large-scale environments. The algorithm is tested in real world experiments with Rolland on the campus of the Universität Bremen. Among others, the so-called "kidnapped robot problem" is solved.

The second topic, the shared-control aspect of manned service robots, is no less challenging. So-called "mode confusion" situations have to be avoided. In the aviation psychology community, the problem of mode confusion has already been discussed for about a good decade. However, the notion as such has never been rigorously defined. In addition, the pertinent publications so far cover almost exclusively the pilot-autopilot interaction.

This thesis presents a new, rigorous view of mode confusion. A framework based on existing formal methods is established for separately modeling the technical system, the user's mental representation of it, and their safety-relevant abstractions. As a result, an automated model checking approach can be applied to detect mode confusion potential already in the design phase. In a case study, the obstacle avoidance skill of Rolland is checked for mode confusion potential with tool support.

This thesis is rounded off with a review on the state of the art in smart wheelchair robotic research and with a brief history of Rolland.

Zusammenfassung

Der Bremer Autonome Rollstuhl "Rolland" ist ein intelligenter Rollstuhl-Roboter, der die Unterstützung behinderter und älterer Menschen zum Ziel hat. Rolland und andere Service-Roboter unterscheiden sich von klassischen Industrierobotern in vielerlei Hinsicht. Die beiden grundlegendsten Eigenschaften sind: Rolland ist mobil und muss häufig in Umgebungen navigieren, die für Menschen und nicht für Roboter ausgelegt sind. Und zweitens: Der Nutzer *und* das technische System steuern Rolland gleichzeitig mit situationsbedingt wechselnden Prioritäten: Rolland ist ein sog. "Shared-Control" System. Zur Umsetzung der geforderten Eigenschaften sind zwei Dinge nötig: Rolland benötigt eine "angemessene" Methode zur Selbstlokalisation und in der Mensch-Roboter Interaktion sind Konfliktsituationen zu vermeiden.

Die Selbstlokalisation mobiler Roboter ist ein lebhaftes Forschungsgebiet. Für kleine und mittlere Umgebungen kann die Aufgabe als gelöst betrachtet werden, sofern eine geeignete Sensorausstattung wie z.B. Laser-Scanner vorhanden ist. Es gibt jedoch noch einige offene Probleme im Bereich der Selbstlokalisation mobiler Roboter. Einige davon sind insbesondere für die Service-Robotik von Interesse: Ziel ist es hier, kostengünstige Lösungen (ein Laser-Scanner für $5000 verdoppelt fast den Preis eines Elektrorollstuhls) zu finden, die auch in weiträumigen Umgebungen, wie z.B. Krankenhaus- oder Universitätsgeländen, funktionieren.

In dieser Arbeit wird RouteLoc präsentiert, ein neues Selbstlokalisationsverfahren, das nur minimale Eingaben und insbesondere *keine* teuren Abstandssensoren benötigt, um einen Roboter in weiträumigen Umgebungen absolut zu lokalisieren. Der Ansatz wird in Experimenten mit Rolland auf dem Campus der Universität Bremen getestet. Das Problem des "entführten Roboters" wird gelöst.

Der zweite Punkt ist der Aspekt der gemeinsamen Steuerung von bemannten Service-Robotern. So genannte "Mode Confusion" Situationen sind zu vermeiden. Im Bereich der Luftfahrtspsychologie wird das "Mode Confusion" Problem bereits seit einem guten Jahrzehnt diskutiert. Trotzdem wurde der Begriff an sich bisher nicht präzise definiert. Die einschlägige Literatur beschäftigt sich bislang ausschließlich mit der Pilot-Autopilot Interaktion.

Diese Arbeit präsentiert eine neue, präzise Sichtweise auf "Mode Confusion". Ein auf existierenden formalen Methoden basierender Rahmen wird vorgeschlagen, um das technische System, dessen mentale Repräsentation durch den Benutzer sowie die sicherheitsrelevante Abstraktion beider getrennt modellieren zu können. Als Ergebnis lässt sich ein automatisches Model-Checking Verfahren anwenden, um "Mode Confusion" Potenzial bereits in der Entwurfsphase zu entdecken. Im Rahmen einer Fallstudie wird Rollands Hindernisvermeidungsverhalten damit werkzeugunterstützt auf "Mode Confusion" Potenzial untersucht.

Diese Arbeit wird abgerundet durch einen Überblick über den Stand der Forschung im Bereich der intelligenten Rollstuhl-Roboter und durch einen kurzen Abriss über die Geschichte Rollands.

Contents

Preface

This thesis is about "Rolland". Rolland is a wheelchair robot. Wheelchair robots are built to support people who are restricted in their personal freedom of action. The goal is to regain as much of the mobility they maybe once had.

In the first place, mobility means to be able to move from your current location to a different one. Obviously, you cannot reliably reach any target without knowing *how* – i.e. along which path – to get there. But in order to know how to get *there*, you have to know *where you are* now. Or to put it the other way round, without knowing its current location a robot cannot plan a path to its goal. Without this plan, it cannot decide which action it should perform in that situation; without knowing where it is, it cannot even decide whether or not it reached the goal. To summarize, being mobile requires knowing where you are.

Mobility has little value if the user is merely automatically transported from A to B. Something is special here in contrast to other robotic applications: the human is "part" of the robotic system. Although he or she may be handicapped, the driver remains an important control unit within the whole system. The user has to be integrated into the navigation process in order to support him or her if need be, but also to exploit his or her (remaining) skills whenever possible. As a consequence, the user has to issue driving commands to the wheelchair robot, which in turn provides the user with some kind of feedback about the current status. Not surprisingly, such interaction scenarios offer much room for misinterpretation between the participants. To summarize, safe and comfortable traveling in a shared-control system requires the human never to be surprised by the robot's behavior.

This thesis covers both aspects: self-localization and shared control. While these topics are scientifically interesting, another, more pragmatic point of view should also be considered when asking for the reasons for wheelchair robot research: wheelchair robots *must* be a booming market in the future. Why is that so? According to Kanowski (2002), Germany is already in fourth place in the

i

world with respect to the average age of its population. Germany is even in third place with respect to the share of people aged 60 years or more. In the future, this trend will even increase. Kanowski (2002) prognosticates a dramatic shift in the age structure. Such a shift seems to be inevitable since the population of modern industry nations ages "from both ends": low birth rates are accompanied by a steadily growing life expectance. As a consequence, the total number of people aged 80 years or more in Germany will increase from about three million today to about five million in 2020 and to an expected number of eight million people in 2050 (11% of the whole population). The effect becomes even more obvious when considering the so-called "Altenquotient" (Kanowski, 2002, p. 27) (i. e. the number of people aged 60 years or more per 100 people between 20 and 59 years of age). This number will increase from about 41 today to around 53 in 2020 and up to about 75 in 2050.

To summarize, the number of elderly people will dramatically increase during the next 50 years. It is rather likely that — despite of all the medical progress — a significant share of those who are 80 or more years old will need some support to maintain their mobility. It will no longer be possible to satisfy the demand for nursing by relatives or even by professional nursing persons. Please note that this will *not* be caused by some "change of values" that led to weaker family bounds, but it will simply be due to the lack of young people who could do the nursing. By further taking into account the scientific and engineering progress in service robotics during the past twenty years, it is not difficult to expect reasonable technical solutions for future home care, transport, and professional nursing applications. As a result, a significant market potential for service and rehabilitation robots in the future seems to be "inevitable" (see also Ch. 1).

There is still a long way to go on the road to a really "intelligent" mobility assistant. However, the fundamental characteristics of such devices have already become clear. Apart from the adaptability to the user's needs and other "soft skills", the key issues when developing a future mobility assistant, will comprise its ability to navigate, the interaction between the user and the technical system, and safety considerations.

As already indicated, this thesis directly covers the navigation aspect as well as the human-robot interaction aspect. The safety concern also plays an important role in Part III.

Organization of this Thesis

This thesis comprises three more or less independent parts.

Part I introduces the context of this work by reviewing the current state of the art in wheelchair robot research, with a special focus on the Bremen Autonomous

Wheelchair "Rolland" in Ch. 2.

Part II presents RouteLoc, a new self-localization approach for mobile robots that works robustly in large-scale environments. This part may either be "browsed" or "studied". For simply getting the idea and basic concepts of RouteLoc, the first three sections of Ch. 4 should be read. A deeper insight into the subject can be gained if the introduction to the subject including a literature review in Ch. 3 and especially the details on RouteLoc in Sect. 4.4 are also studied. In either case, the results of the real world experiments with Rolland, which are presented in Ch. 5, provide the reader with detailed information about RouteLoc's practical performance.

Part III deals with the so-called shared-control problem, which occurs in cooperative situations between human operators and automated systems. After a concise literature review in Ch. 6, a formal framework is proposed that allows to explicitly seperate the technical system, the user's mental model of it, and their safety-relevant abstractions (see Ch. 7). This framework provides the basis for rigorous definitions of the notions "mode", "mode confusion" and "minimal safe mental model" (see Ch. 8). Furthermore, it fosters tool-supported model-checking for mode confusion potential. A case study on Rolland's obstacle avoidance module is presented in Ch. 10.

Contributions

This thesis contributes to the field of mobile service robots on the one hand and to the ongoing debate on mode confusion in the human factors and also in the formal methods communities on the other hand:

RouteLoc. A new absolute self-localization approach for mobile robots is proposed. RouteLoc requires only minimal sensor equipment, works in large-scale environments (indoor/outdoor), and is robust and efficient. RouteLoc as presented here is a fundamental algorithm that is open for various extensions. As far as the author is aware, its "cost-benefit" relation (required hardware, memory, computing time, and sensory bandwidth vs. precision) has not been reached elsewhere so far. The topics are adressed throughout the thesis as listed here:

- Minimal sensor equipment: RouteLoc poses only minimal sensor requirements, as discussed in Sect. 4.2.1

- Large-scale environments: The algorithm is suitable for large indoor and also outdoor environments. The presented experiments were carried out on the campus of the Universität Bremen. See Sect. 4.2.2 and results in Sect. 5

- Efficiency and robustness: ROUTELOC is able to solve tasks such as the "kidnapped robot problem". See Sect. 4.4.6 and results in Sect. 5

- "Cost-benefit relation": See Sect. 5.4 on a discussion about the algorithms "cost-benefit relation", i. e. the required input and resources in relation to the quality of the algorithm's output.

- Open for extensions: The future works section discusses a series of possible extensions that can be added to the approach. See Sect. 5.4.3

Mode Confusion.   The pertinent literature on mode confusion is reviewed. For the first time, rigorous definitions for the notions "mode" and "mode confusion" are given. A formal framework is established that allows to find mode confusion situations by automated, tool-supported model checking. In the first non-aviation case study with respect to mode confusion detection supported by Formal Methods, the obstacle avoidance skill of Rolland is examined. The author's work is the first in the field of mode confusion in (service) robots. The topics are adressed throughout the thesis as listed here:

- Rigorous definitions: Since neither "mode" nor "mode confusion" have been rigorously defined so far, their rigorous definitions may focus research in the emerging direction of "Formal Methods for Human Factors". See Sect. 8.1.

- Formal Framework: The separate modeling of the technical system, the user's mental representation of it, and their safety-relevant abstractions are the foundation of the subsequent automated model checking approach. See Sect. 8.

- Case study wheelchair: With the help of the FDR tool, mode confusion potential has been identified in the obstacle avoidance module of the Bremen Autonoumous Wheelchair "Rolland". See Sect. 10.

Chapter 12 lists 21 publications of the author with their bibliography.

Acknowledgements

Acknowledgements always bear the danger to inadvertently forget to thank some-body who was helpful. In order to avoid such a faux pas, I would like to thank everybody who contributed to this thesis in one way or the other.

However, there are a few persons who I'd like to address personally because without them this thesis would not have been possible in the present form.

First of all I thank my advisor Bernd Krieg-Brückner who always supported me over the years. His attitude of "fördern und fordern" (support and demand) provided me with all the freedom and security needed to do interesting research. I am very grateful that Christian Freksa did not hesitate for a second when I asked him to become my co-reviewer. It was 11 pm during a meeting on the island of Wangerooge...

The Deutsche Forschungsgemeinschaft (DFG) and the Freie Hansestadt Bre-men supported this work by funding the projects "VAMS - Verhaltensbasierte autonome mobile Systeme" (Bremen), "Navigation in Dynamic Environments" (phase II and III) within the DFG priority program "Spatial Cognition" (DFG), and "Formal Fault-Tree Analysis, Specification and Testing of Hybrid Real-Time Systems in Application to Service Robotics (SafeRobotics)" (DFG). I learned a lot when supporting the respective project leaders during the application phase for the two DFG projects.

From a technical point of view, there are two colleagues who helped me more than I could ever expect. I am very grateful to having been given the opportunity to work together with Thomas Röfer and Jan Bredereke. Over the years, Thomas became a mentor and a friend. He was always there when I got stuck with some of the more unpleasant things scientific life offers: last minute late-night preparation of conference talks, finding bit errors in 192 MB large data files, or debugging my ugly code, to mention only a few. To cut a long story short: without him, there would have been no Bremen Autonomous Wheelchair "Rolland". Thanks for all, Thomas. For about a year, I worked with Jan Bredereke who gave me substantial insight into software engineering and system design questions. He helped me with the theory and practical use of the model checker FDR. I thank Jan for this very inspiring cooperation.

Furthermore, I'd like to thank all my current and former colleagues in Bremen, especially the "Mensa gang", for the friendly and always cooperative atmosphere. Robert Ross accepted the "native speaker challenge" and eliminated the remaining Germanisms.

My family always encouraged me to proceed on the long way to the PhD and supported me in different ways: My brother Christian spent an enormous amount of time on thoroughly proof-reading this thesis. My younger brother Bastian never complained about losing his "third son" status with respect to the Bundeswehr as a result of my PhD studies. I very much appreciate that my parents always supported my education without any ifs and buts. I thank my dad for buying the Sharp PC-1245 pocket computer (`10 PRINT "HAPPY BIRTHDAY!"`) and the book "Programmieren ganz einfach" by B. Smith back in 1984. Maybe these two kicked off the whole avalanche, see Fig. 1 taken from this book :-). I thank my mum for her selfless support for all of us during all the years.

Figure 1: *The "Robot Prophecy".*

Above all, I owe a debt of gratitude to my beloved wife Stefanie who suffered a lot from the whole variety of "infinities" that you get for free when writing a PhD: endlessly extended deadlines, neverending night shifts, and the occasionally infinitely bad mood. Nevertheless, she always kept me alive by taking my mind off the worries and thus saved me from becoming a thesis addictive. Kind of "en passant", Stefanie helped me with googling the internet for references, proof-reading and reviewing some of the figures for clarity and expressiveness. Thanks also for keeping me in good shape by always pushing me to run, swim, skate, hike, or play tennis.

I dedicate this thesis to Stefanie and to my parents Ilse and Peter.

Any fault still remaining in this work is my fault.

Part I

Wheelchair Robots

1

Smart Wheelchairs – Overview of an Emerging Market

This review on the ongoing research in the field of smart wheelchair robots is an extended and revised version of Lankenau and Röfer (2000b).

Due to the shift of the age structure in today's industrial populations (cf. the data presented on page i), the demands of the handicapped and the elderly are recognized more and more by politics, industry, and science. The recent development in research areas such as Computer Science, Robotics, Artificial Intelligence, or sensor technology allows significant broadening of the range of possible applications that support handicapped or elderly people in their daily lives. This chapter recapitulates the state of the art of the most popular assistive device: the smart wheelchair.

The increasing mobility of humans will be one of the key issues of this century. People no longer work in the city they live in, huge shopping malls in the open countryside replace the corner shop, and far distance voyages become more and more popular. This phenomenon is accompanied by rapid development, especially in the field of information technology. While notions such as mobile phone, e-commerce, or GPS-navigation system are ubiquitous nowadays, a different application field of the new technologies is still in its infancy: service robotics.

Common industrial robots usually perform repeating movements with enormous speed and precision in an exactly defined environment. In contrast, a service robot's task is to carry out difficult, unpleasant, dangerous or supporting jobs for humans in their normal environment, e. g. surveillance, inspection, cleaning or guidance services.

In a market analysis of October 1999, the United Nations European Commission for Economy (UN/ECE) predicted that the number of service robots installed

3

world-wide would quintuple within three years time (UN/ECE, 1999). This optimism prevailed in the latest UN/ECE study (UN/ECE, 2001). While the market for service robots for professional use (underwater robots, surveillance and cleaning robots) is projected to triple until 2004, that for service robots for private use (cleaning and lawn-mowing robots, entertainment robots) is assumed to increase by more than 500%.

Apart from cleaning robots, the so-called rehabilitation robots, such as smart wheelchairs, play an important role. Such devices make use of the progress of research and development in robotics, computer science, AI, telecommunications, signal processing, sensor technology and other areas such as biology and psychology. By compensating for the specific impairments of each individual, the rehabilitation robots enable handicapped people to live more independently and mobile than before.

This chapter focuses on intelligent wheelchair robots. It summarizes the requirements of a smart wheelchair with regard to the human-machine interface, the technical equipment, functionality and the safety aspect. Furthermore, an overview is presented about current research projects which deal with the development of these vehicles.

In the sequel, the projects that the author is aware of are referred to by their acronyms or project titles, respectively. The following, geographically ordered list may serve as an introductory overview. From North America and Japan, the NavChair (University of Michigan, see e. g. Simpson et al., 1998), the Wheelesley (MIT, see e. g. Yanco, 1998), the Deictic Wheelchair (Northeastern University, see e. g. Crisman and Cleary, 1998), the TinMan (KISS Institute for Practical Robotics, see e. g. Miller, 1998), and the TAO (Applied Artificial Intelligence, Inc., see e. g. Gomi and Griffith, 1998) projects are covered. The European smart wheelchair community is represented by the SIAMO (Spain, see e. g. Bergasa et al., 1999), the CALL Centre (Scotland, see e. g. Odor, 1995), the SENARIO (Greece, see e. g. Katevas et al., 1997), the RobChair (Portugal, see e. g. Pires et al., 1998), the Sharioto (Belgium, see e. g. Nuttin et al., 2002) projects, and the German research groups MAID (University of Ulm, see e. g. Prassler et al., 2001), OMNI (Fernuniversität Hagen, see e. g. Hoyer et al., 1999), INRO (University of Applied Sciences Ravensbrück-Weingarten, see e. g. Schilling and Roth, 1998), and from Universität Bremen the three projects: FRIEND (see e. g. Borgerding et al., 1999), EASY (see e. g. Butler et al., 1998), and Rolland (see e. g. Lankenau and Röfer, 2001a, but mainly Ch. 2).

A chronological order would have shown that the NavChair project was one of the first of its kind. It was installed in 1991. Pursuing an idea of the Director of Physical Rehabilitation at the University of Michigan Hospital, Simon Levine, the group originally intended to apply experiences made in the mobile robotics field to power wheelchairs. For a brief history of the Rolland project please refer to Ch. 2.

Whereas most of the projects are located at universities, there is also one company in the field. In 1995 the Canadian Applied Artificial Intelligence Inc. (AAI Inc.) built TAO, an autonomous wheelchair robot based on the common Canadian platform FORTRESS Model 760V. In 1996, her Japanese sister company presented TAO-2 that was based on the wheelchair Suzuki MC 13-P which is widespread in Japan.

Note that the following review does not pay particular attention to the system architecture of the wheelchair robots. For a systems-oriented review on the architecture and different bus technologies used in smart wheelchairs, please refer to García et al. (2001).

1.1 General Requirements

Three major concerns have to be taken into account when designing a wheelchair robot for handicapped or elderly people: the adaptability to the individual, the fulfillment of safety requirements, and the costs of the add-ons.

1.1.1 Adaptability to the Individual

In order to have a chance of being accepted by its potential users, a smart wheelchair must be adaptable to the needs of each individual person (as an example, cf. the SIAMO project).

Particularly in the context of supporting handicapped people, the focus should be how the remaining skills of the human operator could be complemented adequately. As a consequence, research and industry do not concentrate on fully autonomous systems but on so-called semi-autonomous wheelchairs. These robots are able to carry out certain tasks on their own, but they have to rely on the human operator and his or her skills and experience when performing other tasks. Thus, a smart wheelchair is a highly interactive system, jointly controlled by the human operator and the software of the robot.

As the required interaction between the operator and the wheelchair is not necessarily restricted to simple commands such as "STOP!", the design of the human-machine interface is a key issue in the development of a smart wheelchair. Note that the interface does not only define the input devices to submit commands to the robot but also the output devices of the wheelchair (e. g., voice channel, display).

These considerations imply that the controller and the additional hardware equipment of a smart wheelchair should be organized in a modular architecture. This way, an individual level of support may be provided to each patient. For instance, a young and cognitively healthy paraplegic may only need a collision avoidance module for the back of the vehicle. In contrast, a blind and amnesic user will need a complete coverage by an obstacle avoidance module, and in addition, a global path planner to find certain places in his or her flat.

1.1.2   Safety Requirements

As service robots in general and rehabilitation robots in particular operate in the direct vicinity of humans, their malfunction could cause severe harm to people. Therefore, such robots have to be considered as safety-critical systems (Storey, 1996). For smart wheelchairs, this classification is even more reasonable because they transport persons who often depend completely on the correct behavior of the technical system. If, e. g., the handicapped operator of the wheelchair instructs the vehicle to go to the medicine cabinet, the dependable execution (Laprie, 1992) of the command might be life-critical, failure would not be an option.

Whereas it is commonly established that power plants, aircraft and railway systems are safety-critical, the majority of research groups dealing with smart wheelchairs do not pay very much attention to this topic. Only few groups invest a lot of effort into the question how to design a safe smart wheelchair. For instance, the Rolland project aims at applying formal methods such as hazard analysis techniques (Lankenau et al., 1998) and model checking to define safety requirements of the system (Lankenau and Meyer, 1999), to prove the satisfaction of these requirements during operation and to tackle the so-called mode confusion problem that arises in shared-control systems (see Part III of this thesis). The importance of the safety issue becomes obvious when considering that it is the most prominent fear of industrial service robot manufacturers to cause a severe accident and thus to bury the whole market before it could develop.

1.1.3   Costs

As smart wheelchairs will be purchased either by private people or by health insurance companies for the insured, the additional costs caused by the sensor equipment, the human-machine interface, and the computer hardware should be rather moderate. Since 1993, the KISS Institute for Practical Robotics has been selling robotic wheelchairs (TinMan) to universities at cost, in order to push smart wheelchair research.

Many of the projects being presented in the sequel kept this in mind when choosing the hardware for their wheelchair robots. Most of them use expensive laser range finders only for research purposes (such as self-localization), the cheap infrared or sonar sensor technologies are preferred when developing market-oriented applications (such as safe traveling).

1.2 Functionality

The variety of required functionality is as large as the amount of different handicaps. The realization of the necessary skills must be easy to use by persons who do not have a technical education. A smart wheelchair has to work reliably and robustly in the natural environment of its user. It is not acceptable that this environment must be completely rebuilt in order to let the wheelchair operate as intended by the developer. Nevertheless, there is some very recent effort (e. g. by Kantor and Singh, 2002) to make changes to the environment affordable by using extremely cheap active landmark systems (see also the discussion of this topic on page 49). Maintenance and configuration of such wheelchairs have to be as intuitive as possible because they should be carried out by the staff of the reha-provider, and not by the robotic expert.

In the following subsections, a brief overview of the relevant skills figured out so far is given. In addition, the research projects mentioned above are classified with regard to the topics they work on.

1.2.1 Obstacle Detection

On the one hand, the quality of the detection of obstacles is a question of the sensor equipment used. On the other hand, it is a question of the interpretation, representation, and processing of the data provided by the sensors. For instance, a horizontally mounted infrared sensor can never detect a pothole in front of the robot, but maintaining a local occupancy grid drastically reduces the probability that sonar sensors miss an obstacle (Röfer and Lankenau, 1999b).

Sensor Equipment

Every project tracks the locomotion of the vehicle by processing the current speed and the direction of movement delivered either by externally mounted wheel encoders or by the internal wheelchair electronics. In contrast, the employed proximity sensors vary significantly. Sonar sensors are very common. Often, these are mounted in a ring around the wheelchair (e. g. Rolland, SENARIO), but sometimes they only cover the front of the vehicle (e. g. NavChair, INRO). The SIAMO project developed a special sensor setup that avoids sonar cross-talks (Ureña et al., 1999). Bank (2002) presents a sonar sensor system for mobile robots (not explicitly mounted on a wheelchair, but easily adaptable) that is able to detect multiple echos from different echo paths for each sensor.

Infrared sensors are also fairly widespread (e. g. RobChair, Wheelesley, and SIAMO), but only the TAO project uses them as the only active proximity sensors. As they are relatively expensive, laser range finders, very popular in the mobile robotics community, are only rarely used (e. g. MAid). Some projects use laser range finders when doing research with their wheelchair robots but do not intend to employ them in the market version.

Prominent passive proximity sensors are bumpers providing a binary signal whether or not they are in touch with an obstacle (e. g. Deictic, Wheelesley, and TAO). These tactile sensors are quite reliable but suffer from the fact that if a bumper touches an object, there is only a minimal period of time left before the collision of the wheelchair with the obstacle will occur. The Deictic project uses a special bumper which is of soft foam and thick enough to ensure that the wheelchair will stop in time even when traveling at maximum speed (Crisman and Cleary, 1998). Other passive sensors are video cameras also used to estimate distances to objects in the surroundings, e. g. in a stereo vision system (TAO, Deictic) or by exploiting optical flow. The Deictic project mounted the stereo camera system on a pan-tilt unit in order to broaden the range of this sensor. Cameras are additionally used to detect potholes or descending staircases by determining the deviation of the actual shape of a laser beam from the target shape in the picture (INRO, Senario).

Handling of Sensor Measurements

Apart from the choice of sensor hardware, one has to decide how to use the data provided by the system. As almost every project implemented a basic safety layer to avert collisions, the primary purpose of the proximity sensors is to allow the control software to stop in time when an obstacle is dangerously close to the wheelchair. Only the TAO robots employ a direct sensor-action coupling and do not store the data delivered by their sensor system. The majority of the other projects maintains a local obstacle map (or "certainty grid") to accumulate sensor readings. The idea of this approach is to cope with sensor misreadings such as sonar crosstalks or specular reflections. In addition, such a local map allows the update rate of each single sensor to be reduced without losing too much information. Each sensor reading acquired is stored in such a map. In addition, the map is shifted according to the wheelchair's motion in reality, employing its dead reckoning system. Thus, the position of the wheelchair relative to an obstacle in the map always represents the corresponding distance in reality (NavChair, Senario, Rolland).

1.2.2   Obstacle Avoidance

To ensure safe traveling, a smart wheelchair has to provide a reliable obstacle avoidance skill. However, there are various interpretations of the notion "obstacle avoidance" among the projects.

Reactive Obstacle Avoidance

The purely reactive approach is exclusively promoted by the TAO project: the current sensor readings are directly mapped to motor actions. If the human operator does not accept a decision of the so-called TAO Autonomy Management System, he or she is able to override the command with a contradicting joystick command. This idea is in a direct contrast to the approaches pursued by other smart wheelchair projects: the TAO-wheelchair uses the common control interface, the joystick, to explicitly taking control and circumventing all safety measurements. This seems to be problematic for certain user groups, e. g., if the patient suffers from spasms from time to time, it is rather likely that he or she hits the joystick in an uncontrolled fashion. As the joystick commands override the decision of the Autonomy Management System, a collision would happen.

Obstacle Avoidance Based on a Local Map

The most popular obstacle avoidance approach is the use of a local obstacle map. By accumulating the most recent sensor readings, a rather reliable detection of potential obstacle is ensured. Note that the term "recent" varies from project to project, but it is a widespread technique to "forget" old sensor readings after a while in order to avoid the map being flooded with former misreadings. Really existing obstacles will be detected again and again, so they are not forgotten.

The most prominent approach is used by the NavChair project. The original intention of this group was to apply the promising results of mobile robot obstacle avoidance by the so-called Vector Field Histogram method to a power wheelchair (Borenstein and Koren, 1991). It soon turned out that maneuvering a power wheelchair is more complex than a common mobile robot. Therefore, the NavChair now employs the Minimum Vector Field Histogram (MVFH) method. MFVH consists of four steps: First, the sensor readings are inserted into a Cartesian local obstacle map which extends around the wheelchair. This map is regularly updated and shifted such that the wheelchair is always located in the center of the map. Each cell in the grid map represents an area in the real world whose obstacle status is represented by a number that counts how often a sensor detected an obstacle. Second, the obstacle map is transformed into a polar histogram, the angular component of which represents the travel direction. The amplitudes of the histogram show the obstacle density. Third, a weighting function is added to the polar histogram. It is a parabola with its minimum at the indicated travel direction. As the minimum of the weighted polar histogram determines the chosen travel direction, the weighting function is a parameter that can be changed in order to adjust the behavior of the wheelchair in certain situations. Finally, the target speed of the wheelchair is calculated according to the distance to the closest obstacle in the intended travel direction. This method finds a compromise between the user's goal direction and the best (with respect to the expected collision-free travel distance) possible direction. Thus, e. g., the wheelchair is guided smoothly through doorways. Other research groups (e. g. Senario, Sharioto) use various extensions to the vector field histogram method.

In the Rolland project, the obstacle avoidance skill is implemented differently. The wheelchair is always located in the center of a Cartesian local obstacle map, i. e. if it moves, the whole map is shifted translaterally. As rotational shifts are computationally expensive, the orientation of the wheelchair within the map is modified instead. This map is the basis for the extensive use of the principle of function tabulation: On the one hand, for each combination of travel direction, steering angle, and orientation of the wheelchair in the map, the cells of the grid that the wheelchair would visit if it moved are calculated in advance and stored as so-called virtual sensors. As a consequence, at any time the question of whether or not an intended movement is dangerous can be answered very quickly. On the other hand, it is pre-calculated for each cell in the map, how fast the wheelchair must be at most to pass an obstacle located in that cell. The third component of the obstacle avoidance skill is the user's intention. The travel direction indicated by the joystick is projected into the local obstacle map. If the closest dangerous obstacle lies on the left of the projection, the algorithm decides to detour the object on the right, and vice versa. If, instead, the projection of the travel direction indicated by the human operator points directly towards the obstacle, it is assumed that the user does not want to detour the object. This is very helpful in situations in which the driver wants to approach a cupboard closely. Note that a detailed description of Rolland's obstacle avoidance approach can be found in Sect. 6.2 in Part III of this thesis.

1.2.3   Solutions to the Shared-Control Problem

The problem of shared control always arises when a human operator and a technical system are jointly in charge of control. As an example consider the co-existence of pilot and auto-pilot in today's aircraft. As many accidents in safety-critical systems were due to this problem, a recently established research community deals with the development of general solutions to questions such as how to model shared-control systems and how to avoid catastrophic consequences if human operator and technical system are not aware of the other ones present state. Please refer to the detailed discussion on mode confusion in wheelchair robots in Part III of this thesis.

Meanwhile, the smart wheelchair projects had to find practical realizations for the restricted application of controlling a semi-autonomous mobile robot. The obstacle avoidance approaches used in the NavChair and in the Rolland project pay attention to the shared-control problem in that they consider the human operator's intention where to travel as the bias direction.

By providing a set of disjoint operating modes, NavChair is able to adapt to specific situations. These are the General Obstacle Avoidance Mode, the Door Passage Mode, and the Automatic Wall Following Mode. Two different methods of automatic mode transitions are combined with the help of a Bayesian network to ensure that the systems chooses the correct mode in a certain situation. These implicit mode transitions are somewhat problematic, because—due to a lack of awareness—the user might miss a transition and misinterpret subsequent actions of the wheelchair.

The Deictic project implements another form of shared control. If an obstacle is detected, the wheelchair stops. It is the user's task to decide how to circumvent the object. He or she decides on the basis of a video image of the environment, which of the possible actions will be chosen. The wheelchair's task is to convert the abstract actions such as "Pass the object on the left-hand side with strolling speed" into adequate motor commands.

1.2.4   Behavior-Based Skills

The human-machine interfaces used in the smart wheelchair projects (cf. Sect. 1.2.6) enable the user to instruct the robot on a significantly more abstract level than an operator of a common power wheelchair could do. Thus, several projects implemented various local navigation skills such as corridor following, object tracking, turning on the spot, and others. The approaches to passing through doorways have already been described in the obstacle avoidance section (cf. Sect. 1.2.2). The TAO wheelchairs use the stereo camera system in combination with the infrared sensors to detect corridors or walls which can be followed. Its behavior-based Autonomy Management System (AMS) ensures that at any point in time one of the hierarchically organized behaviors is the active behavior. If a corridor is detected, the AMS switches to "centering in the corridor" mode. As mentioned above, the human operator of the TAO wheelchair is able to regain control if he or she sets a driving command via the joystick.

In addition to the widespread behaviors such as wall and corridor following, Rolland has a turning-on-the-spot skill. This is necessary, because the Meyra Genius 1.522 wheelchair is a non-holonomic vehicle driven by its front wheels and steered by its back wheels. Moving forwards with this kind of wheelchair is like driving backwards in a car. In order to waste as few space as possible during the turning maneuver, the basic idea is to split the required 180° turn into portions of preferably 90 degrees (in case there is no obstacle around). This algorithm also works in the presence of obstacles in the environment of the wheelchair. Then, a sequence of smaller segments of circles is combined. The algorithm automatically choses an appropriate direction in which to turn, based on the absence of obstacles.

In contrast to the Rolland wheelchair, the experimental platform of the OMNI project uses four so-called Mecanum wheels. These wheels are able to move with three degrees of freedom in the plane. Therefore, turning on the spot is trivial for this vehicle. As mentioned above, the Deictic wheelchair allows the human operator to issue rather abstract commands which refer to objects in the surroundings of the robot. The commands are executed by tracking the relevant object in the video picture taken by the stereo vision system. After accomplishing a subtask, the wheelchair stops and waits for new user input.

1.2.5   Navigation

The task of large-scale navigation is difficult to realize in the context of smart wheelchairs. Their inherent purpose is to increase the mobility of their users. As a consequence, a navigation approach should not rely too much on the environment, i. e. it should not be necessary to modify the environment in order to let the navigation approach work. For instance, the common induction loop technique used for robots in factories has to be replaced by intelligent methods such as using already existing features of the environment, e. g. to follow corridors. Nevertheless, the basic requirement of navigation is a reliable self-localization technique. It is a key challenge for the research groups in this area to provide self-localization methods working in natural environments that are not necessarily known in advance.

A popular approach for the wheelchair to adapt to various scenarios is learning by tuition. After service staff has trained the wheelchair to operate in a certain environment, it is able to perform navigation tasks in that environment. During the training process the system has to build a map of its environment which is subsequently matched with the real world using a self-localization technique. Among the projects presented here, there are some that employ topological maps (TAO) and others that use a combination of topological and metrical maps (Rolland, Senario).

In the Senario project, global path-planning is done with the help of a qualitative map. It is based on a topological diagram of the environment that is segmented to a grid of rectangular cells each of which represents one node of the topological diagram. In a simulation, a qualitative sensor impression is calculated for each cell. In addition, the relative variation of each sensor reading in a cell with respect to the corresponding reading in a neighboring cell is stored. These data make up the qualitative map. If the target of the robot is known, it is possible to determine a preferred travel direction for each cell of the qualitative map by processing the topological relations between the cells. The online path planner matches the real world position of the wheelchair with the qualitative map and chooses the preferred travel direction.

For outdoor navigation, the satellite-based Global Positioning System GPS may be employed. The INRO project makes use of this technique for self-localization by a differential GPS module.

1.2.6 Human-Machine Interface

The smart wheelchair projects can be divided into two groups with respect to the question whether or not they do research about designing a suitable human-machine interface. Those groups that confine themselves to other problems such as navigation or the safety issue include NavChair, Rolland, EASY, INRO, and others. They simply use the standard joystick as input device and provide no special output device apart from simple displays. Some groups (RobChair, SIAMO) employ speech recognition systems to enable the user to issue commands by voice. As the recognition rate of the available standard speech recognition systems is still not convincing, it seems not advisable to rely on this technique only. In order to reduce this problem, both projects only allow a small set of commands that can be recognized rather reliably. The disadvantage of this approach is that in an emergency case a patient might be inclined to use a command that is not known to the system.

In the Wheelesley project, the only means for human-machine interaction is a Macintosh Powerbook placed on a table in front of the user. The human operator controls the wheelchair by choosing high-level commands via a graphical user interface on the notebook. There are two input devices: a single switch system and the so-called EagleEyes technology, which tracks the eyes' movements with electrodes (Yanco, 1998). When using the single switch method, the display shows a set of icons standing for possible actions in the corresponding situation. Each icon, in sequence, is highlighted for a while. With the help of the single switch, the user can select the highlighted command to become the active command. Alternatively, the selection may be carried out with the EagleEyes system. By processing the information received from electrodes that are fixed to the user's skin around his or her eyes, a cursor can be moved over the display. This technique is especially useful for people who cannot control their arms and hands. Nevertheless, it is relatively time consuming and requires some experience.

The SIAMO project provides even more input devices: Apart from joystick control, switches and a voice recognition system, it offers a blow control and a facility that enables the user to instruct the wheelchair by head movements. This system mainly consists of a CCD micro camera mounted in front of the user in order to track his or her face. After a skin segmentation, the eyes and the mouth are located in the picture by extracting the hollows in the skin blob. In combination with the face tracking, this method yields enough information to determine head movements or facial expressions such as a turn to the left, upward movement, winking eyes, etc. The features recognized are matched to motor commands: turning the head to the left results in a left turn of the wheelchair. Apart from minor difficulties, various test persons were able to control the wheelchair with this system after a short training.

In order to control the Deictic wheelchair, the human operator has to use a four component control panel as human-machine interface. He or she can choose a motion, a direction, the placement of the most important object close to the wheelchair, and a distance or speed. This input device has successfully been tested in simulation and reality. The robot is called *deictic* since the user controls the wheelchair by telling the robot how to move relative to the important objects he or she pointed out in a video image.

The experimental platform of the FRIEND project is equipped with a control-PC and a robotic arm structure, the MANUS manipulator. The main topics of the project are the control of the manipulator and its human-machine interface (Martens et al., 2001). As a result, there are two ways of using the MANUS arm: replaying preprogrammed movements in known environments and user control by speech recognition.

The INRO as well as the RobChair project employ a radio link from the wheelchair to a remote station. On the one hand, this link can be used by the service or nursing staff to tele-operate the vehicle, e. g. if neither the user nor the wheelchair itself find a way to further proceed in a difficult situation. On the other hand, the wheelchair is able to permanently transmit its current state (including its position) to the remote station. Both projects did not yet cope with the control of "intelligent" devices in the environment such as elevators or automatic doors via the radio link.

1.2.7   Special Hardware Features

In the OMNI project, the seat of the wheelchair can be elevated up to 90cm. Thus, the user is very flexible with regard to omnidirectional movements on the floor as well as with regard to his or her vertical level of operation (e. g. for taking a book from the shelf). The FRIEND project tries to realize this functionality with the help of their MANUS manipulator.

1.3   Expected Developments for the Future

In spite of the convincing progress the smart wheelchair community made within a good ten years of its existence, there is still a lot of work to do before such devices will become commercially available. The research wheelchairs are not yet robust enough to operate for a long time in the house or flat of a handicapped person. In order to increase the acceptance in the potential buyers' mind as well as to ease official certification, the safety issue has to be examined more thoroughly. Nevertheless, the chances to provide a useful tool to significantly improve the quality of life for many people are quite realistic. Maybe that's why the company of Johnson&Johnson invested $50 million to develop IBOT, a smart wheelchair that is able to climb stairs (ACM, 2000). IBOT was projected to be commercially available by the end of 2001, at about $25,000. However, it is not yet on the market.

16

2

*Genius is one percent inspiration
and 99 percent transpiration.*
Thomas Edison

A Brief History of "Rolland"

This chapter introduces the Bremen Autonomous Wheelchair "Rolland". The name "Rolland" was coined by Bernd Krieg-Brückner. It is a combination of the german word "Rollstuhl" for wheelchair and the name of the symbol of the independence and liberty of the Freie Hansestadt Bremen, the "Roland" monument.

Strictly speaking, Rolland is the second prototype of the Bremen Autonomous Wheelchair. The first one was set-up in the student project SAUS. Several publications use it as an experimental platform (Kollmann et al., 1997; Röfer, 1997a,b, 1998a; Röfer and Müller, 1998). Today, the first prototype is no longer in use.

Rolland serves as an experimental platform in Part II, and it is the subject of the case study in Part III of this thesis. Therefore, the hardware, the system architecture, research activities and applications of the wheelchair are briefly recapitulated here. This chapter closes with an overview of Rolland's "evolution".

2.1 Wheelchair and Hardware

The Bremen Autonomous Wheelchair "Rolland" (cf. Fig. 2.1) is based on the commercial power wheelchair Genius 1.522 manufactured by the German company Meyra. The wheelchair is a non-holonomic vehicle driven by its front axle and steered by its rear axle. The human operator controls the system with a joystick. The wheelchair has been extended by a standard PC (Pentium III 600MHz, 128 MB RAM) for control and user-wheelchair interaction tasks, 27 sonar sensors, and, recently, a laser range sensor behind the seat. The sonars are arranged around the wheelchair such that they cover the whole surroundings. The electronics is able to simultaneously fire two sensors, one on the left side and one on the right side of the wheelchair. An intelligent adaptive firing strategy has been introduced by Röfer and Lankenau (1999a,b). The sonars are mainly used for obstacle

Figure 2.1: The *Bremen Autonomous Wheelchair "Rolland"*. Photo by Rolf Müller.

detection. The laser range finder has an opening angle of 180° towards the back-side of the wheelchair and is able to deliver 361 distance measurements every 30 ms. The laser range finder is used for mapping and localization purposes. The original Meyra wheelchair already provides two serial ports to set target values for the speed and the steering angle as well as determining their actual values. Data acquired via this interface is used for dead reckoning. The odometry system based on these measurements is not very precise, i.e. it performs relatively well in reckoning distances but it is weak in tracking angular changes. See also Fig. 4.15 on page 69.

2.2   Architecture

A modular hardware and software architecture (see Fig. 2.2 on the next page) based on the real-time operating system QNX allows the adaptation of Rolland to an individual user (Röfer and Lankenau, 2000). This section describes the system architecture from an implementation point of view, whereas the following section focuses on the corresponding research topics. The various software modules com-

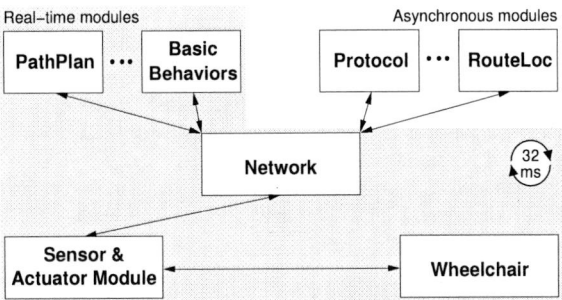

Figure 2.2: *System architecture of the Bremen Autonomous Wheelchair "Rolland".*

municate via a real-time network with each other. The idea of the network protocol by Lankenau and Meyer (1997, 1999) is that non real-time applications such as a complex image processing module can never block the real-time modules such as the sensors and actuators module (SAM). SAM provides a safe interface to the physical wheelchair: a higher level module (or the user) may give an arbitrary driving command. If the obstacle situation around the wheelchair requires to do so, SAM changes this command to a safe one. On the other hand, SAM provides the higher level modules with all the information delivered by the sensors of the wheelchair.

Figure 2.2 shows the set-up as currently used on the wheelchair. All software modules run on the same control-PC as separate QNX processes. As a consequence, the network module operates as a black board relay station that receives data from and provides data to the communicating modules. However, the architecture lends itself to extension to more than one control-PC.

The user controls the commercial version (no control PC, no sensors) of the wheelchair with a joystick. The command set via the joystick determines the speed and the steering angle of the wheelchair. The idea of Rolland's safety module (Röfer and Lankenau, 2000) is to wiretap the control line from the joystick to the motor. Only those commands that will not do any harm to the wheelchair and its user are passed unchanged. If there is an obstacle dangerously close to the wheelchair, the so-called *safety module* performs an emergency brake by setting the target speed to zero. The notion "dangerously" refers to a situation in which there is an object in the surroundings of the wheelchair that would be hit, if the vehicle was not decelerated to a standstill immediately. Thus, this fundamental module ensures safe traveling in that it guarantees that the wheelchair will never actively collide with an obstacle.

Above the safety module, higher-level skills provide additional functionality: obstacle avoidance (i. e. smoothly detouring around objects in the path of the wheelchair), assistance for passing the doorway, behavior-based traveling (wall following, turning on the spot, etc.) and others. These modules have been combined to the *driving assistant* (see below). It provides the driver with various levels of support for speed control and for steering.

The research activities group around Rolland can be classified in the four categories simulation, navigation, formals methods, and applications:

2.3   Research Activities

Rolland has to date been used as the experimental platform in more than 30 scientific robotics and formal methods publications so far. The majority of this work was conducted by Bernd Krieg-Brückner's working group at the Universität Bremen. In addition visiting researchers (e. g., see Marsland et al., 2001) or working groups from other German universities (e. g., see Musto, 2000) used Rolland for various kinds of experiments. Furthermore, its presentation at public venues such as the Hannover Messe Industrie trade show in 1999 attracted a lot of media coverage (Röfer and Lankenau).

A workshop co-organized by the author dealt with the topic of safety in service-robotics applications such as Rolland (Lankenau and Röfer, 2000; Röfer et al., 2000).

Most of the work referred to here was and still is supported by the Deutsche Forschungsgemeinschaft (DFG). Rolland continues to play a key role in the DFG funded projects "Navigation in Dynamic Environments" within the DFG priority program "Spatial Cognition" (1997-2003), "Formal Fault-Tree Analysis, Specification and Testing of Hybrid Real-Time Systems in Application to Service Robotics (SafeRobotics)" (2001-2004), and "Automatic Diagnosis of Strategies of Other Mobile Robots in a Cooperative, Dynamic Environment" within the DFG priority program "Cooperating Teams of Mobile Robots in Dynamic Environments" (2000-2002, may be extended).

2.3.1   Simulation

The kinematic robot simulation tool SimRobot (Röfer, 1998c; Siems et al., 1994) allows us to develop and debug software for the wheelchair in "virtual reality". New control modules can be interchangeably used either on the physical wheelchair or in the simulated version of it. Another interesting feature of the simulation is the replay mode: SimRobot can use real world data that was recorded by Rolland while traveling to exactly replay the scenario of the real world. This turned out to be extremely useful especially during the development of the self-localization algorithm ROUTELOC which is presented in Part II of this thesis (see also Sect. 5.1.1 on page 100).

2.3.2 Navigation

While developing Rolland's navigation skills, contributions were made with respect to modeling navigation knowledge, local navigation, route navigation, self-localization and map building. These topics are briefly discussed here. Röfer and Lankenau (2002) recapitulate the route-related results of the "Navigation in Dynamic Environments" project.

Modeling Navigation Knowledge

Werner et al. (1997) present a taxonomy of navigation which identifies three classes of navigation knowledge: location knowledge, route knowledge, and survey knowledge. Routes are defined as sequences of locations which in turn are specified as being characterized by a view of the surrounding at a given position. Werner et al. (2000) integrate such routes and so-called places to route graphs (see also the discussion in Sect. 4.2.2 on page 49).

Local Navigation

Rolland's local navigation skills enable the wheelchair to safely operate in typical environments such as office buildings, campus areas, hospitals, etc. Besides the fundamental safety functionality provided by the "Sensors and Actuators Module" (SAM), the local skills comprise shared-control obstacle avoidance (Lankenau and Röfer, 2001a; Röfer and Lankenau, 1999a,b), turning-on-the-spot (Lankenau and Röfer, 2001a), and various *basic behaviors* such as wall-following, corridor-following, and turning-into-door (Krieg-Brückner et al., 1998; Röfer and Lankenau, 2000; Röfer and Müller, 1998).

Route Navigation

Extending the local navigation skills to navigation along routes provides a first simple means of traveling from a point **A** to a remote point **B** in a given environment. A route as understood by Werner et al. (2000) is a one-dimensional data structure. Navigation along such routes requires being able to self-localize within a route. Röfer's (1998b; 1999b) work on the generalization of odometry data to abstract route descriptions turned out to be a powerful alternative to the previously used landmark-based approach (Röfer, 1997b; Röfer and Müller, 1998). Müller et al. (2000) employ coarse qualitative route descriptions usually used by humans to control a wheelchair along a route.

When traveling, Rolland is able to extract maps of the environment by dividing the route traveled so far into a sequence of straight segments that join under certain angles. By incrementally generalizing these descriptions, a kind of canonical representation of the routes is obtained (see also Sect. 4.2.1). Self-localization in these route segments is simply done by dead reckoning. There are two applications of these route representations implemented so far. The so-called route assistant, and a behavior-based route navigation approach. The route assistant is discussed below. The behavior-based route navigation approach additionally attaches information about the currently active behavior to these route descriptions (Röfer, 1999b). If the wheelchair is autonomously replaying such a stored route, it can invoke the adequate behavior at the correct position on a specific route. This approach is quite robust with respect to dynamic environments, because it does not fail to replay a route if there is an obstacle in its way which has to be avoided before the route can be finished.

Self-Localization and Map Building

The navigation along single routes is not enough if application scenarios such as patient transport within hospitals or the like are taken into account. Especially situations in which the robot erroneously leaves a route are not handable in this case. Therefore, the self-localization capability of Rolland had to be extended to maps that may cover whole large-scale environments. Two approaches are pursued here: the first is a probabilistic approach for the self-localization of the robot in a pre-existing topological-metric representation of the environment. This work is the subject of Part II of this thesis. Preliminary versions have already been published elsewhere (Lankenau and Röfer, 2001a; Lankenau and Röfer, 2002; Lankenau et al., 2002). The second approach allows the robot to simultaneously self-localize and build a map of its environment. It employs a histogram-based correlation method to relate consecutively recorded laser scans. The algorithm developed works in real-time, and is able to detect loops and keeps the map built consistent (Kollmann and Röfer, 2000; Röfer, 2001a,b, 2002).

2.3.3   Formal Methods

Decreasing mental and physical abilities of users of service and especially rehabilitation robots such as Rolland lead to an increasing dependence of the human on the technical system. Therefore, these assistive devices should be regarded as safety-critical systems, such as aircraft or power plants. As a result, formal methods, commonly accepted to be state of the art in the safety-critical systems community, have to be transferred to the service robotics domain. Using Rolland as a demonstrator, two aspects have been examined so far: hazard analysis and mode confusion. In a fault-tree based hazard analysis, a detailed safety requirements specification for the Bremen Autonomous Wheelchair was defined (Lankenau and Meyer, 1999; Lankenau et al., 1998). Furthermore, the satisfaction of some safety requirements was verified with the model checking tool FDR (Roscoe, 1997). The approach provides an additional specification of the environment in which the robot is able to operate safely.

On the other hand, the experiences with mode confusion problems in the aviation domain are also relevant in rehabilitation robotics applications. Therefore, the behavior of the obstacle avoidance module of the wheelchair was formally specified in CSP. Subsequently, it was compared to the mental model of a potential wheelchair user by means of a refinement check with the model checking tool (Lankenau, 2001). This work has been substantially enhanced by Bredereke and Lankenau (2002). Rigorous definitions of the notions mode and mode confusion are given, three interfaces are identified that play an important role for mode confusions in shared-control systems, and a new classification of mode confusions by cause is presented which leads to a number of recommendations on how to avoid mode confusions when designing shared-control systems. The author worked on these topics within the DFG funded project SAFEROBOTICS. Part III of this thesis deals with the mode confusion problem.

2.3.4 Application Scenarios

At the moment, the two main applications implemented for Rolland are the *Driving Assistant* and the *Route Assistant*. In the future, the *Navigation Assistant* will provide a global navigation module.

Driving Assistant

The *Driving Assistant* (Lankenau et al., 1998; Lankenau and Röfer, 2001a; Röfer and Lankenau, 2000) provides an interface of a safe wheelchair to higher level navigation modules or the user, respectively. The principal idea is to wiretap the connection between the joystick and the motor. If the human operator issues a command that may lead to a collision with an obstacle, the wheelchair autonomously changes the dangerous command into a safe one. To be able to make this decision, the Driving Assistant maintains an occupancy grid as local obstacle map. The 27 sonars are ring-wise mounted around the vehicle. The sensors deliver distance information that is stored and processed in the map. The Driving Assistant provides higher level modules with a variety of useful operations: turning-on-the-spot, collision detection, and obstacle avoidance. The obstacle avoidance approach is presented in Sect. 6.2 in Part III of this thesis.

Route Assistant

The *Route Assistant* of the Bremen Autonomous Wheelchair (Lankenau and Röfer, 2001a) has been developed in cooperation with the neurological clinic of a Bremen hospital. It provides the following functionality: During a teaching phase, the system explores the routes and places pertinent for the future user(s). If, e. g., the wheelchair is used in a rehabilitation center for amnesic patients, the routes to all relevant places in the building could be learned and stored for later replay with the help of the generalization algorithm mentioned in Sect. 4.2. In the replay mode, a nurse chooses a certain target for the patient in the wheelchair. Similar to a GPS-based navigation system, the large-scale navigation is done by the Route Assistant by giving instructions where to go at decision points, enabling the patient to travel around on his or her own. The patient is independently responsible for controlling the vehicle with respect to local maneuvers such as obstacle avoidance. At the current state of development, the Route Assistant is restricted to scenarios in which it initially knows its position in a specific route. Then, it is able to direct the user to the goal of this single route. Since the Route Assistant is intended for patients, e. g. amnesic people, in a hospital environment, this solution seems to be sufficient since the starting positions are inherently known (e. g., the patient's room). Although it is adequate for following single routes, the current implementation of the Route Assistant is not very robust because the user may inadvertently miss an instruction by the wheelchair and subsequently leave the route. Even though the departure of the wheelchair from the route can be detected, up to now it has not been possible to recover from the error and return to the route.

Navigation Assistant

The integration of the Route Assistant and self-localization techniques such as RouteLoc, will be named "Navigation Assistant". The Navigation Assistant is going to solve large-scale navigation tasks with Rolland in the near future.

2.4   Summarizing Overview

In order to visualize the main topics of Rolland related research during the past five years, Fig. 2.3 shows the "evolution" of the Bremen Autonomous Wheelchair "Rolland". The chart is organized along the two dimensions "wheelchair

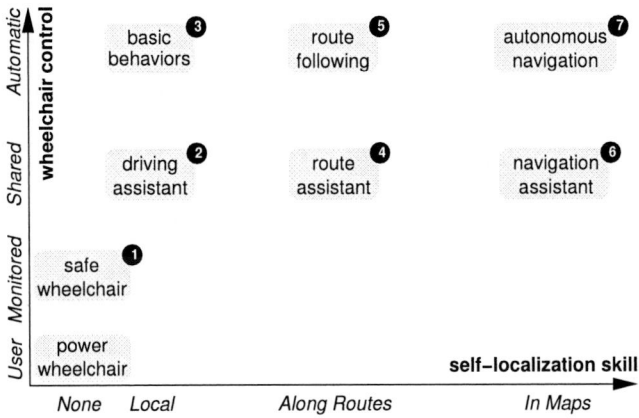

Figure 2.3: *Evolution of the Skills of the Bremen Autonomous Wheelchair.*

control" (vertical) and "self-localization skill" (horizontal). The grey-shaded labels mark the main milestones in Rolland's evolution from a commerical off-the-shelf power wheelchair (bottom left; the user is in control, the wheelchair has no self-localization ability) to a smart wheelchair robot that is able to autonomously navigate in large-scale environments (top right; wheelchair drives autonomously, it can find its position in a map of the environment) .

The following list shows the relevant publications that are closely related to the milestones visualized in Fig. 2.3. The bullet numbers in the list correspond to those in the figure. Please note that "related to" does not necessarily mean that, for instance, Röfer (2002) describes the Navigation Assistant. However this paper will have provided some brick that helped to build the Navigation Assistant.

❶ Lankenau and Meyer (1997, 1999); Lankenau et al. (1998); Röfer and Lankenau (1998); Röfer and Lankenau (2000)

❷ For the robotic aspect of shared-control please refer to Lankenau and Röfer (2001a); Röfer and Lankenau (1999a,b). For the formal aspect of shared-control please refer to Bredereke and Lankenau (2002); Lankenau (2001); Lankenau and Röfer (2000).

❸ Röfer (1997b, 1998b, 1999b); Röfer and Lankenau (2000); Röfer and Müller (1998)

❹ Lankenau and Röfer (2000a, 2001a)

❺ Müller et al. (2000); Musto et al. (1999); Röfer (1999a); Röfer and Müller (1998)

❻ Lankenau and Röfer (2001b); Lankenau and Röfer (2002); Lankenau et al. (2002); Röfer (2001a,b, 2002); Röfer and Lankenau (2002); Werner et al. (2000)

❼ Subject of current research, no publication yet.

Part II

Self-Localization in Large-Scale Environments for the Bremen Autonomous Wheelchair "Rolland"

28

3

Near Potsdam, [...] a 57 years old driver and his car [...] fell into the Havel river. [...] the navigation system was not aware of the ferry in Caputh: The driver simply followed the road — straight into the water.
translated from: Deutsche Presse Agentur, December 26th, 1998

Mobile Robot Self-Localization

This thesis part embeds the field of self-localization in the broader context of mobile robot navigation in Sect. 3.1. Then, a review on mobile robot self-localization is given in Sect. 3.2. In Sect. 3.3, general requirements for a self-localization approach for service robots are specified. Finally, a new self-localization approach in large-scale environments is presented in detail in Ch. 4 (the ROUTELOC algorithm) and Ch. 5 (experimental results).

3.1 Navigation — a Brief Introduction

The notion "navigation" has become a buzzword in recent years: Portable GPS receivers for satellite based navigation are sold in supermarkets for less than $150, the free-ware Internet browser "Netscape Navigator" substantially contributed to the success of the world wide web, the University of Michigan hosts a "Knowledge Navigation Center" providing a "campus gateway to the knowledge resources of the Library and the world", middle-class automobiles have "car navigation systems" as a standard fitting that are able to guide the driver through whole continents, and so on.

But what does *navigation* mean literally? Navigation is of Latin origin. It is a concatenation of *navis* (for "ship") and *agere* (for "to act, to do, to drive"). The original, rather narrow, meaning of the word has encountered some proliferation as the definitions in Encyclopædia Britannica and Merriam-Webster Collegiate Dictionary show:

> NAVIGATION: *science of directing a craft by determining its position, course, and distance traveled. Navigation is concerned with finding the way, avoiding collision, conserving fuel, and meeting schedules.*
>
> *(Encyclopædia Britannica)*

> NAVIGATION: *the science of getting ships, aircraft, or spacecraft from place to place; especially: the method of determining position, course, and distance traveled.*
>
> *(Merriam-Webster Collegiate Dictionary)*

Interestingly, according to both definitions neither animals nor humans are considered to be able to navigate. However, they agree upon the fact that "determining the position" of the agent is important for navigation. This view is supported by a more technical and meanwhile classical definition given by Gallistel (1990):

> NAVIGATION *is the process of determining and maintaining a course or trajectory from one place to another. Processes for estimating one's position with respect to the known world are fundamental for it. The known world is composed of the surfaces whose locations relative to one another are represented on a map.*

A similar view is taken by Levitt and Lawton (1990) who define navigation as the process of answering the three questions "Where am I?", "Where are other places relative to me?", and "How do I get to other places from here?". The same questions are put by Leonard and Durrant-Whyte (1991). Since the early nineties, many publications on mobile robot navigation refer to either of both papers to substantiate the claim that self-localization is indispensable for robot navigation.

Some authors find the questions too restrictive and call for changes. For instance, Trullier et al. (1997) propose to formulate the questions in a more general fashion. As an example, consider the first question "Where am I?". It can only be answered if you have some representation (e. g., a map) of the environment. Then, the answer specifies your position within the map. Trullier et al. (1997) prefer not to implicitly require the a priori existence of a map-like environment representation. Therefore, they propose to replace the "Where am I?" question by "What are the identifiable characteristics of this place?".

Nearly all publications on mobile robot navigation follow one or the other of the above definitions. Nevertheless, ethological research found evidence that in human and animal navigation the question "Where am I?" is not necessarily the first one of the three to raise. And even more, some animals are able to navigate without answering any of the three questions, i. e. they find their goal without self-localization.

Franz and Mallot (2000) use this as a point to generalize the aforementioned definition by Gallistel (1990). In their concise review on navigation in biomimetic robots, they no longer demand Gallistel's requirements for self-localization capability and map-like representation of the environment.

Instead, they define

> NAVIGATION *is the process of determining and maintaining a course or trajectory to a goal location.*

Based on this definition, Franz and Mallot (2000) specify a hierarchy of navigation techniques some of which indeed do not require self-localizing the agent. For instance, the navigation technique "searching" finds a goal only by chance during some exploration of the environment. According to Franz and Mallot (2000), the hierarchical navigation techniques can generally be classified into two categories *local navigation* and *way finding.*

Please note that if no self-localization is available, the animal, the human or the robot has to be able to decide whether or not the goal is reached by some direct perception. This maybe easy for animals and especially for humans with their huge sensor capacity and world knowledge, but is often difficult for robots. In this context, Kuipers and Beeson (2002) introduce the notion *image variability* for a location that looks differently on different occasions. This is the converse of *perceptual aliasing* as discussed in Sect. 4.3.4 on page 68.

Therefore, the definition by Franz and Mallot (2000) may be most convincing when animal, human, and robot navigation should be covered in a common framework. But if the focus is solely on (service) robots one should be more explicit with respect to the self-localization capability: It may be true that animals exist that have the sensor capability of perceiving external stimuli that allow them to find their goal without explicit self-localization. But it is definitely also true that self-localization in some map-like representation is an indispensable prerequisite of robot navigation if a robot lacks this sensor capability. To make it more explicit: A robot such as Rolland could not even decide whether or not it reached the goal in situations that offer low sensor information if it had no self-localization skill.

As a result of this discussion, it seems reasonable to consider self-localization to be a fundamental skill for mobile robot navigation. This is even more important in large-scale environments where searching may take too much time.

Or, as Cox (1991) puts it:

> *... to locate the robot in its environment is the most fundamental problem to providing a mobile robot with autonomous capabilities.*

After having pointed out the importance of self-localization for mobile robot navigation, the question has to be answered, in what kind of environment the robot has to self-localize. Obviously, the more complex (i. e. the larger, the less feature-rich, etc.) the environment, the more difficult is the localization task. The claim of Franz and Mallot (2000) that even a compass guided journey over thousands of kilometers has to be classified as *local navigation* and does not require the agent to recognize places other than the goal itself is not very convincing.

In order to describe the complexity of the robot's environment, it is often referred to its "scale". Wehner (1996) uses a categorization of animal navigation in small-scale, middle-scale, and large-scale environments. Wehner (1999) examines the impressive large-scale navigation performance of desert ants which is mainly based on path integration but supplemented by other techniques such as landmark guidance.

Duckett (2000) transfers this terminology from biology to robotics. He refers to small-scale navigation if the robot does not leave the lab. He refers to middle-scale navigation if the robot leaves the lab and enters "unmodified public area such as corridors...". Finally, he refers to large-scale navigation if the robot "must leave indoor environments and navigate over much larger distances".

Since the self-localization approach RouteLoc, which is introduced in this thesis, claims to be suited for large-scale environment, this category is discussed here in some more detail:

Duckett's definition of large-scale navigation has drawbacks: Firstly, it requires that the robot must operate outdoors. But this is definitely no distinctive property for large-scale, because huge airport buildings or trade show areas reach the size of small cities and should be classified as large-scale environments even though they lack the outdoor component. Secondly, mentioning "larger distances" is not useful, because it remains unclear how large is "larger". And, third, there is no relation to the size of the robot, its perceptual range, its operational range or the like.

Trullier et al. (1997) are more precise and define "large-scale environment" as an environment "in which there are relevant cues out of the range of perception, and in particular where the goal is not in the immediate environment". The "range of perception" is a number relative to the robot. The navigation in large airport buildings would correctly be classified as large-scale here.

In Kuipers (1977), the notion "large-scale" is defined relatively by referring to large-scale space if the environment cannot be perceived at once. As a consequence, a small but cluttered room is regarded as large-scale whereas a large city could be considered as small-scale if seen from an airplane. This view appears to be counter-intuitive, even though it is consequent. Later, Kuipers and Byun (1991) phrase it more specifically when characterizing an environment as large-scale "if its spatial structure is at a significantly larger scale than the sensory horizon of the observer." This is again close to Trullier et al. (1997) and seems to be a reasonable demarcation to middle-scale or small-scale navigation. To eventually avoid the confusion with the "small-scale city", the following extension suggests itself:

An environment is considered as large-scale *if its spatial structure is at a significantly larger scale than the sensory horizon of the observer who/that is operating within the environment.*

After having clarified the notions used in the headline of this part of the thesis, the following section gives a concise literature review on self-localization.

3.2 Literature Review

This chapter briefly recapitulates the state of the art in the self-localization of mobile agents (robots, vehicles, people). Since this is a field of rapid development due to ongoing world-wide research activities, it seems reasonable to say that the relevant literature up to and including the *International Conference on Robotics and Automation* in May 2002 has been taken into account for this overview (cf. ICRA'02).

3.2.1 Robot Self-Localization

There are several criteria regarding how the existing approaches could be structured. The characteristics of the RouteLoc algorithm presented in this thesis are its model of the robot's environment and the model of the robot's situation. Apart from RouteLoc almost all of the following self-localization approaches make extensive use of proximity sensors to define their situation. For this reason the following categorization uses the representation of the environment of the different approaches as distinctive feature. The categories are presented in the three following sections: metric (see Sect. 3.2.1); topological (see Sect. 3.2.1); and hybrid topological-metric approaches (see Sect. 3.2.1). As RouteLoc is intended to self-localize a mobile robot in a network of corridors, it is not that far away from localizing a car in a network of streets. Therefore, Sect. 3.2.2 gives a concise insight into the localization approaches known for automotives.

Please note that this overview refers only to the self-localization aspect of the cited approaches, even though work on Simultaneous Localization and Map Building (or *SLAM*, cf., e. g., Choset and Nagatani, 2001) as well as Localization and Planning (cf., e. g. Simmons and Koenig, 1995) is included. Please note that, as a complement to this overview, Sect. 5.4.2 provides a detailed comparison between RouteLoc and a small number of other prominent self-localization approaches.

Independent of the way of modeling the environment, there are two basic principles for the self-localization of mobile robots (Borenstein et al., 1996): *Relative* approaches need to know at least roughly where the robot started and are subsequently able to track its locomotion. At any point in time, they know the relative movement of the robot with respect to its initial position, and can calculate the robot's current position in the environment. It has to be ensured that the localization does not lose track, because there is no way to recover from a failure for these approaches. Modern relative self-localization methods often use laser range finders. They determine the robot's locomotion by matching consecutive laser-scans and deriving their mutual shift. Gutmann and Nebel (1997) and Gutmann et al. (2001) use direct correlations in their *LineMatch* algorithm, Mojaev and Zell (1998) employ a grid map as "short term memory", and Röfer (2001a) accumulates histograms as basic data structure for the correlation process.

On the other hand, *absolute* self-localization approaches are able to find the robot in a given map without having any a priori knowledge about its initial position. More impressively, they solve the "kidnapped robot problem" (Engelson and McDermott, 1992), where – during runtime – the robot is deported to a different place without being notified. From there, it has to (re-)localize itself. That means, the robot has to deliberately "unlearn" acquired knowledge. The absolute approaches are more powerful than the relative ones and superior in terms of fault tolerance and robustness. But so far they had to rely on exteroceptive information, GPS signals, laser range finder data or the like. Absolute approaches usually try to match the current situation of the robot – defined by its locomotion and the sensor impressions – with a given representation of the environment, e. g. a metric map.

The following sections summarize the current state of the art in self-localization in the categories introduced above. For an extensive review of the state of the art then, please refer to Thrun (1998).

Metric Localization

As the matching problem is intractable in general, probabilistic approaches have been proposed as a heuristics. The idea is to pose a hypothesis about the current position of the robot in a discrete model of the world from which its location in the real world can be inferred. A distribution function that assigns a certain probability to every possible position of the robot is adapted stepwise. The adaptation depends on the locomotion performed and the sensor impressions perceived by the robot. Due to the lack of a closed expression for the distribution function, it has to be approximated. One appropriate model is provided by grid-based Markov-localization approaches that have been examined for some time: They either use sonars (Elfes, 1991) or laser range finders (Burgard et al., 1997) to create a probability grid. As a result, a hypothesis about the current position of the robot can be inferred from that grid. Arras and Tomatis (1999) show that the integration of monocular vision and a laser range finder is able to improve the localization results in comparison to single sensor systems. These methods are commonly referred to as "Markov" localization approaches, because they exploit the so-called Markov condition, see also Sect. 5.4.2. Recently, so-called Monte-Carlo-localization approaches became very popular. They use particle filters to approximate the distribution function (Fox et al., 1999; Thrun et al., 2000a). As a consequence, the complexity of the localization task is significantly reduced. Nevertheless, it is not yet known how well these approaches scale up to larger environments.

Topological Localization

Purely topological environment representations are by orders of magnitude less complex than metric maps. But they also lack the precision realizable with a metric representation. However, there were and there still are several interesting approaches that solve the localization with adequate precision in purely topological maps. Since there exists no distance information in topological approaches, the only means to decide where the robot is, is place recognition.

Hertzberg and Kirchner (1996) present a self-localization approach for robots operating in sewerage pipes. The method is similar to Simmons and Koenig (1995) but does not use any metric information. Place recognition is solved by detecting landmarks with a rotating ultrasound transducer that allows to measure distances up to 1 m. The sonar image is classified by a neural network to relate it to certain predefined landmark categories.

Gutierrez-Osuna and Luo (1995) use a state-set progression method similar to the one introduced by Nourbakhsh et al. (1995) to navigate in a purely topological environment representation. Place recognition is done with sonar sensor images.

An interesting alternative for the distance sensor based place-recognition are vision oriented methods. For instance, Röfer (1995, 1997a, 1998a) implements a navigation approach for the first prototype of the Bremen Autonomous Wheelchair using 360° panoramic images. In a teaching phase, the robot learns routes as sequences of such images. In a replay phase these routes can be followed autonomously by driving from one key image to the next. In a similar approach, Ulrich and Nourbakhsh (2000) use a purely topological representation of the environment: an adjacency graph the nodes of which represent locations (e. g., rooms in an apartment) and the arcs represent the adjacency relationship between them. They use a passive color vision camera that delivers panoramic images as the only sensor. During a learning phase, reference images for relevant locations are stored. In later runs, the robot uses a histogram-based matching approach to find the best match for its current impression. In indoor and outdoor experiments Ulrich and Nourbakhsh (2000) found a correct classification rate between 87 and 98 percent.

Hybrid Topological-Metric Localization

Apart from these purely metric or topological representations of the environment, Kuipers et al. (1993) propose the integration of metric and topological concepts with their "spatial semantic hierarchy" (see also Kuipers, 2000). The idea is pursued by Simmons and Koenig (1995) by augmenting topological maps with metric information. The resulting self-localization methods also work probabilistically on the basis of the odometry and a local model of the environment perceived by the sensors. Details about these two prominent approaches are given in comparison with RouteLoc in Sect. 5.4.2 on page 126.

Cassandra et al. (1996) use a Bayesian model to represent the robots environment by a set of states. The topology of these states is defined by a set of actions that execute transitions between the states. In their experiments, each state represents an area of 1 m × 1 m and one of the four compass directions. The approach adapts from a uniform probability distribution over the set of states to estimate the state corresponding to the current robot position. Each state is assigned with certain actions that allow the transition to an adjacent state. For instance, a state representing a region in front of a doorway could be left by a door-passage behavior.

Choset and Nagatani (2001) use generalized Voronoï graphs to encode the topology and also some metric information of the robot's environment. Low-level control laws enable the robot to explore unknown territory by generating and following the Voronoï graph edges. Localization is done by matching (local) graph fragments with the previously constructed global graph. Due to the nature of Voronoï graphs (sets of points equidistant to two obstacles in the plane), Choset and Nagatani (2001) report that their approach works well in environments which are rich in topological information, i. e. cluttered with obstacles. The benefit of using the representation of the environment itself as perceived by sensors as "landmarks" has to be paid for with two drawbacks: On the one hand, slight changes in the environment (e. g., in public buildings crowded with moving people) result in matching problems. On the other hand, open environments do not provide enough information at all to construct a Voronoï graph.

Zwynsvoorde et al. (2000, 2001) pursue a similar approach to that presented by Choset and Nagatani (2001). They also infer topological Voronoï-like representations from local sensor perceptions and merge them with a global graph. Whereas Zwynsvoorde et al. (2000) confine themselves to map building, Zwynsvoorde et al. (2001) also cover the localization task.

Tomatis (2001) and Tomatis et al. (2001a,b, 2002) present a hybrid grid-based and topological approach. They employ a 360° laser range finder and extract features such as corners and openings which are used to navigate in a global topological map. The feature extraction is based on earlier work (Arras and Tomatis, 1999). In addition, the laser-scans are searched for line structures (walls, cupboards, etc.) which build the basic data structure for several local metric maps (one for each node of the topological map). Tomatis et al. (2001a) find a success rate of 100% after the first successful localization in 25 experiments that amount to a total travel distance of about 900 m. But at transition points, they report some 1.4% false state estimates.

3.2.2 Localization for Dedicated Applications

Since the late 1980s more and more applications for everyday situations have emerged which require a localization component: mobile phones, car navigation systems, personal navigation systems, and recently, the promising market of location-based services. This section briefly covers the most prominent approaches for localizing cars on digital street maps (in the following subsection) and mentions mobile navigation systems for personal digital assistants (see below).

Localization of Car Navigation Systems

Apart from psychologists, biologists, geographers, roboticists and computer scientists, there is another group of researchers that deal with self-localization of mobile vehicles: automotive engineers. Browsing the literature on robot self-localization results in the finding that the robotics community is very well aware of work done by biologists and psychologists with respect to animal and human navigation, but they seem to more or less ignore the car navigation and intelligent transportation systems community. A reason might be that most robotic research so far has been bound to navigation in either *structured* indoor environments or *unstructured* outdoor environments. Nevertheless, some of the work done for car navigation systems is highly relevant for future robot developments. Apart from the human-machine interaction aspect which has been and still is examined for cars and has to be so for manned service robots (see Part III of this thesis), the navigation and self-localization approaches might also be used in some special robot applications.

In his book on the state of the art in vehicle location and navigation systems, Zhao (1997) considers car self-localization to comprise a *positioning* and a *map-matching module*. The positioning module is used to determine the vehicle's position with respect to some reference coordinate system. Available sensor sources are, e. g., dead reckoning, gyroscopes, and the Global Positioning System (GPS). While the pure position information would be satisfying for some robotics applications (e. g., the robocup soccer scenario), car navigation requires the relation of the real world position of the vehicle to a location represented in a digital street map. Finding such a relation is referred to as *map-matching* in the intelligent transportation systems community. According to Zhao (1997), the different map-matching approaches used in vehicle location systems are:

Semi-Deterministic Approaches. Pre-requiring that the initial location and the travel direction of the vehicle is given, so-called *semi-deterministic* algorithms use the digital street map to correct the odometry calculated position during travel. If the projection of the odometry position onto the digital street map leaves the street, the position estimate is corrected. This approach fails immediately if the map is inaccurate (road missing, constructions, etc.), the car leaves the road network represented in the map, the tracking is accidentally lost, or if the car is temporarily transported by other means, for example by train. Then, no system-immanent recovery is possible.

Probabilistic Approaches. If the odometry is processed probabilistically, it is no longer necessary to define that the road network must not be left. At program start, the initially known starting position, which is by definition on the route network, hosts the whole probability mass: In each update step, the position estimate based on the dead reckoning data is filtered with an uncertainty component. Then, the route segments that intersect the area in which the car could be are searched in the digital street map. Based on a set of factors such as heading, connectivity and closeness, a candidate segment is determined. Similar to the semi-deterministic approaches, this information is used to correct the odometry-based position. If the algorithm is in doubt, because more than one candidate segment appears plausible, no correction takes place.

Fuzzy-Logic-Based Approaches. Determining a hypothesis about which route segment hosts the car can also be done with the help of fuzzy-logic-based algorithms. By specifying a set of fuzzy rules that are processed in each update step, it is possible to track the vehicle's trajectory through the network of corridors. For instance, a rule could state that a change of the heading direction of 180° means that the possibility of a U-turn is high. Depending on the set of rules, such an approach is able to solve the problem of steadily tracking the vehicle's position. Nevertheless, it is not able to deal with temporarily matching losses, or the kidnapped robot problem.

The interesting part of car self-localization is that it is a real *large-scale* application. The industry pushes the development (see, e. g., Bachmann and Bujnoch, 1999) since there is an enormous market potential in these systems. A market study presented in Richardson and Green (2000) predicts a growth of the market volume in the US by an order of magnitude until the year 2015; by the year 2009 more than a quarter of all cars in the USA will be equipped with navigation systems. As a result, very precise digital street maps of whole continents have been developed. Recently the International Organization for Standardization has begun an initiative to take the map formats from Europe (Geographic Data Files, GDF), Japan (Japan Digital Road Map Association, JDRMA) and the United States (Spatial Data Transfer Standard, SPTS) and unify them into a common standard (ISO/TC 204).

In general, car navigation systems have to provide the following functionality: The driver is able to somehow choose the intended trip destination (for a report on how this should be done, cf. Nowakowski et al., 2000). Then, in a loop that terminates if the target is reached, the system estimates the position of the car in a digital street map, plans a path to the target regarding some constraints, and instructs the driver if he or she has to change the travel direction.

From the robotic point of view, the drawback of all approaches presented in this subsection is that they are relative or at most "semi-absolute" techniques. That means, if they once lose track of the position, they require an external input (e. g. a new GPS signal, or user input) to re-localize them. When CARIN, the **car** information and navigation system developed by Philips Research since 1984, was still in its infancy, the driver had to enter his or her starting position manually. And, after using car trains, ferries, or passing longer tunnel passages, the driver had to once again re-enter the position of the car. This was due to the fact that in the late 1980s, the GPS was not yet implemented in CARIN (cf. (van Oosterom, 1991, page 15f)). Thus, no source of absolute information was available to those systems, apart from the user input. Strictly speaking, the localization components in early car navigation systems were only able to *track* an originally known position while traveling. They were explicitly not able to solve the kidnapped robot problem as introduced above.

With the usage of external data sources such as the GPS signal or the cell information from the GSM network, "semi-absolute" techniques became applicable. They provide almost the same efficiency as true global approaches, as they use the external stimuli as a first approximation of the real car position and employ various techniques to fuse this with dead reckoning data to be able to issue a precise position estimate. As a result, these systems use the external position signal to define a "search window" that defines a map area in which they perform an absolute localization more or less reliably.

Even though these systems are working in today's cars, research is still searching for the best technique capable of localizing a car in a digital street map. Four recent approaches with four different matching methods are gathered here:

Kim and Kim (2001) implement a map-matching algorithm that is based on adaptive fuzzy networks. The worst-case localization error in a real world experiment has been found to be below 15 m.

Using the Monte Carlo Localization technique (MCL) known from the fields of mobile robot self-localization (see above) seems reasonable. In two very recent publications, MCL is employed to determine the position of a vehicle in a network of roads (Forssell et al., 2002; Gustafsson et al., 2002). Still requiring a rough estimate of the initial position, the approach is then able to track the vehicle's location. The only input to the algorithm are the initial position, current odometry data and the information delivered by the digital street map.

Czommer (2001) examines different algorithms for the correlation of the perceived car and target trajectory defined by the digital street map. All of them are based on the assumption that the initial mapping between real world position and map location is known.

Lamb and Thiébaux (1999) present a two-step localization approach that avoids the direct map-matching and is similar to the one by Simmons and Koenig (1995). It uses a Markov model to maintain a set of hypotheses regarding which road segment most probably hosts the car at a time. Within each road segment, a Kalman filter is used to predict the car's position within the segment. Since Lamb and Thiébaux do not obtain the position input from a dead reckoning system but from an absolute sensor, namely a roadside beacon system, the presented experiments in a simulated and in a real road network only show the behavior of the approach when exposed to bounded errors. Then, the mean location error during a trip in a 200 m × 300 m area is 9 m.

Bernstein and Kornhauser (1996) briefly examine different map-matching algorithms that use geometric as well as topological information about the environment. They confine themselves to relative localization as they assume the initial position to be known.

Since all these approaches rely on the occasional re-calibration by the GPS information, it is relevant to ask whether a GPS update always improves the position estimate. Abbott and Powell (1999) find that the performance of today's car navigation systems with respect to the error in the position estimates is dominated by the accuracy of the GPS information when the signal is available. In situations where no GPS signal is available, the rate gyroscope's bias drift dominates the error.

An interesting "human factors" aspect of car navigation systems is discussed by Rogers et al. (1999). They present an interactive agent for route advice that adapts to user profiles.

Localization of Mobile Navigation Systems

Due to the increasing success of mobile devices such as personal digital assistants (PDAs) or the ubiquitous cellular phones, applications making use of so-called location based services become more and more interesting both for academic as well as for industrial research. For instance, Baus et al. (2002) present a hybrid navigation system that uses different localization techniques (e. g., GPS for outdoor tasks and infrared beacons for indoor scenarios). The presentation of the current situation and the route instruction is adapted to the limited technical resources of the user interface. Furthermore, the limited cognitive resources of the user are paid attention to. Experiments were conducted indoors in the Computer Science Department of Saarbrücken University as well as outdoors on the campus of Saarbrücken University.

3.3 Self-Localization of Service Robots – The Requirements

From the state of the art in self-localization, some — to a certain extent generic — requirements can be derived. If a manned service robot fulfills these requirements, the self-localization task could be considered as solved for this specific device.

Absolute. As should have become clear, only absolute self-localization approaches are robust enough to solve the difficult task in the service robotics context. The task is difficult, because service robots such as surveillance, cleaning or assistive robots often have to operate in environments that are made for humans and *not* for robots. As a consequence, they are dynamic, potential features may disappear, etc. Therefore, it is difficult to identify those environment features that do not change too often. Otherwise they are not suitable as landmarks.

Large-scale. Since a lot of application fields of service robots are large-scale environments, the self-localization approach should be able to determine the robot in an environment of that size. For instance, consider a mail-delivery robot. This application becomes economically interesting if it can serve at least a complete building.

Low-cost. In order to minimize costs, a good self-localization approach should require little of anything: little sensor requirements, little computing capability, little or no changes to the environment, etc.

Open. A good self-localization approach should be open for extensions (e. g., the integration of additional sensors). Furthermore, it should be transferable to other domains, such as using robot self-localization approaches in car navigation systems.

"Human Compatible". Especially in the context of manned service robots, it is important that the self-localization algorithm delivers a result that can be understood by the user. To illustrate the difference, try to imagine where the location 2°17′47″ of East longitude and 48°52′24″ of North latitude is. If you are not sure, this is probably not because you don't know the location. Further more, it is likely that you have already been there. If you cannot imagine where this location is, it is because the description in latitude/longitude coordinates is definitely not "human compatible". Better location descriptions for this location from a human point of view are "addresses" or verbal descriptions such as *L'Avenue des*

a)   b)

Figure 3.1: *"Human Compatible" Self-Localization.* Photos by the author.

Champs Élysées, at the Place de l'Étoile close to the Arc de Triomphe in Paris, France or even pictures or sketches, see Fig. 3.1. But it is also important that the location specification contains relevant cues that are known to the human: The view in Fig. 3.1a could have been taken in many big cities worldwide. The photo in Fig 3.1b was taken at the same location, with the orientation changed by 180°. It is far more expressive. This is because the depicted monument is a cue known to most people. Such cues are rare in technical location representations such as the longitude/latitude position.

Therefore, a good self-localization algorithm for manned service robots should provide its position estimate in a format that is not only unambiguous for the algorithm but also for the human.

4

When it is not in our power to determine what is true,
we ought to follow what is most probable.
René Descartes

ROUTELOC: Self-Localization in Large-Scale Environments

This chapter evolved from prior joint work with Thomas Röfer published in Lankenau and Röfer (2001b); Lankenau and Röfer (2002). A short version of a preliminary version of this chapter has also been published as Lankenau et al. (2002). Thomas Röfer also helped coding earlier versions of the implementation.

This chapter presents the first major contribution of this thesis in detail: ROUTELOC. ROUTELOC is a self-localization algorithm for mobile service robots operating in large-scale environments.

4.1 Introduction

As discussed in Ch. 1, service and rehabilitation robots are considered to play an important role in the future development of assitive devices for the home care sector as well as for professional usage in hospitals and old-age homes. Future service robots are likely to be required to offer flexible task solving strategies for a variety of applications. Being flexible very often means being mobile, and being mobile means being able to move from A to B to navigate. As shown in Sect. 3.1, navigation requires self-localization. Therefore, a service robot's mobility depends on its self-localization competence.

But what kind of self-localization approach is adequate for service robotic applications? Service robotic applications need an approach that is in agreement with the requirements asked for in Sect. 3.3. This chapter presents a new self-localization approach that satisfies these requirements: ROUTELOC. ROUTELOC works robustly in structured large-scale environments; ROUTELOC is a basic approach that lends itself to extensions in several ways; already the basic version is able to absolutely localize a robot in a large-scale environment; furthermore, ROUTELOC copes with the "kidnapped robot problem".

As input data, ROUTELOC requires a topological-metric map and a so-called incremental route generalization. It calculates a position estimate in a human-compatible form, e. g. "The robot is in the corridor leading from **A** to **B**, about x meters past **A**".

The strengths of the approach are its minimal resource needs (with respect to computation time, memory, and sensor equipment) and its scalability to large environments. As ROUTELOC, in the presented version, offers some prospects for further development, its current drawbacks are likely to be overcome in the future.

In the following section, the models for the situation of the robot and for the environment are introduced. Section 4.3 presents the basic ideas of ROUTELOC. Technical details and a deeper insight are given in Sect. 4.4. To validate the algorithm, experiments with the Bremen Autonomous Wheelchair were conducted. They are recapitulated in the results chapter, Ch. 5, which also proposes some ideas for future work in this field (Sect. 5.4.3).

4.2   Modeling Locomotion and Environment

As mentioned above, self-localization of mobile robots is often achieved by matching the robot's situation (i. e. the current and past sensor impressions and its locomotion) with a representation of its environment. "Matching" means to find a correlation between two models, the situation model and the environment model.

Please note that the notions "environment model" and "situation model" are commonly used in the pertinent literature *without* being defined consistently. While the usage of environment model is rather similar throughout the publications, some irritations with respect to "situation model" occur. For instance, Duckett (2000) uses a "location model" instead. His location model does however relate the robot's situation to a reference in the environment model.

4.2.1   Situation Model

In this thesis, the "situation model" is a description of the current situation of the robot based only on information available to the robot, i. e. dead-reckoning and sensor readings. Explicitly, it does not contain the correlation to the real world. For example, a natural approach is to snapshot the robot's configuration at a time t to represent its situation.

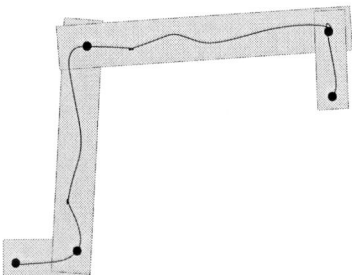

Figure 4.1: *Route Generalization (Röfer, 1999b).*

Two disjoint classes of situation models exist: sensor-based and action-based. If the representation of the robot's situation is sensor-based, the robot might use the whole set of sensor readings at a certain time to specify its situation. Due to sensor noise, some filters are likely to be applied here to group ranges of sensor readings to discrete values. As for action-based situation models, the sensor readings are ignored except those from the odometry system. Nothing but the information delivered by the robot's odometry is used to describe its situation.

Route Generalization

The approach follows on from an idea of Musto et al. (1999), Röfer (1999b) which introduces an incremental generalization of traveled tracks. The idea is to generalize the locomotion of a traveling robot at runtime to an abstract route description. Such a description represents the route as a sequence of straight segments that intersect at certain angles. Since natural minor deviations which occur while traveling are abstracted away, the generalized description of the robot's route from its starting point to its current location is an adequate situation model. Please note that there is a significant difference between the incremental route generalization as introduced by Röfer (1999b) and standard line-reduction algorithms, such as e. g., the "original" by Douglas and Peucker (1973). The incremental route generalization works on incomplete knowledge since the route shape is not fixed while the robot travels. Nevertheless, the underlying geometric ideas are quite similar.

Specifying Abstract Route Descriptions. Figure 4.1 shows the locomotion of the robot, recorded by its odometry system, as a solid curved line. The corners recognized by the generalization algorithm are depicted as circles. The rectangular boxes represent the so-called acceptance areas. As long as the robot remains within such a region, it is assumed that the robot is still located in the same corridor. The width of the rectangular boxes is determined with the help of a histogram-based approach from the measurements of two sonars mounted on the wheelchair's left- and right-hand side chassis (Röfer, 1999b). Note that there may be other generalization algorithms that do not rely on external sensor input.

As a result, the generalization R of the route traveled so far is defined as a sequence of *corners* as follows:

$$R = \langle c_i \rangle, \text{ where } c_i = (\rho_i, l_i), \ i \in \{0, \dots, n\} \quad (4.1)$$

In contrast to the concept "corner$_c$" proposed by Eschenbach et al. (1998), the length of the incoming segment of a corner is not considered here. Thus, in (4.1), ρ_i is the rotation angle between the incoming and the outgoing segment of a corner in a "local frame of reference", i.e. ρ_i describes the relative change in orientation when passing corner c_i. As an example, consider the almost rectangular corner c_1 in the lower left part of Fig. 4.1 (c_0 is the "virtual" starting corner): ρ_1 is about 86°, because the robot has to turn about 86° to the left when changing corridors at corner c_1. Note that ρ_0 is a "don't care" value, i.e. only the outgoing segment of the first corner is considered, whereas the angle is ignored. The second parameter of a corner as specified in (4.1) is the length l_i of its outgoing segment.

Other definitions of the concept "route" in the pertinent literature (Hunt and Waller, 1999; Kuipers, 2000; Werner et al., 2000) only slightly differ in details such as terminology and notation, but usually agree on the basic idea that a route is a sequence of decision points that are connected by segments. Depending on the problem domain, this general definition can be instantiated for different real world scenarios. In their approach to the formalization of navigation, Remolina and Kuipers (2002) rigorously define a route as a *path connecting regions*. A "region" is a set of places. A "path" defines an order relation among places that are connected by travel with no turn actions. And finally, a "place" is a set of distinctive states linked by turn actions.

Incremental Generalization of Route Descriptions.   Since the situation of the robot has to be known while it travels, the route generalization must be carried out incrementally and in real-time. Röfer's approach satisfies both requirements. Nevertheless, the incremental generalization has the drawback that it partially has to rely on uncertain knowledge: the distance l_n already traveled in the current segment as well as the angle ρ_n to the previous segment may change at runtime depending on the locomotion of the robot. The information about c_n is volatile and not fixed before a new final corner c_{n+1} is detected. This is illustrated in Fig. 4.2 on the facing page.

The upper row of the figure shows three different snapshots of a single trajectory driven by the robot. The respective current location of the robot is indicated

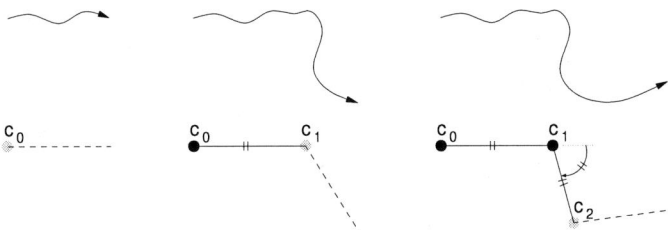

Figure 4.2: *Fixing of the Penultimate Corner During the Incremental Generalization.*

by the arrow. Even though this is only a sketch, it is reasonable to expect a similar odometry recording when the robot travels in a straight corridor, turns right after some time and turns left some time later. In the lower row, the corresponding generalizations are shown: In the leftmost column, no corner has been detected so far, the traveled path completely fits into the imaginary corridor defined by the acceptance area of the segment depicted as a dashed line. In the middle column, the robot has conducted a right turn and already seems to be performing a new turn to the left. Nevertheless, it is only then that the robot leaves the acceptance area of the first segment. As a result, the generalization algorithm sets up a new—so far final—corner (indicated by the grey circle) and a new—also so far final—segment (indicated by the dashed line). Simultaneously, the parameters of the first corner c_0 (now marked by the black circle) are fixed. Since c_0 is the first corner, the angle is irrelevant; but the length of the outgoing segment is now known. In the right column of the figure, the robot has moved further and has left the acceptance area of the second route segment, resulting in the set up of another new segment. The generalization algorithm positions the third corner and fixes the parameters of c_1: the rotation angle from the first to the second segment and the distance between c_1 and c_2.

The abstraction resulting from this generalization method turns out to be very robust with regard to temporary obstacles and minor changes in the environment. Nevertheless, it is only helpful, if the routes are mainly connected as networks of corridors or the like. Fortunately, almost all larger buildings such as hospitals, administration or office buildings consist of a network of corridors. In such environments, the presented algorithm works robustly.

Extero- vs. Proprioception A substantial advantage of relying only on the locomotion data is that the definition of the current situation is independent of "environmental noise". Here, it is no problem, if you cannot detect any feature in the surroundings to use as a landmark, people strolling around do not interfere with sensor readings, and so on. Please refer to Sect. 3.2.1 on page 35 for a discussion of the finding that environmental noise is a problem for Voronoï based approaches (Choset and Nagatani, 2001).

Further note that this assessment is a direct contradiction to Duckett's thesis that self-localization on the basis of proprioceptive data is "unsuitable" (Duckett, 2000, p. 27). Duckett substantiates his claim by stating that the proprioceptive robot could not re-establish the correct position estimate after it becomes lost, because it cannot rely on any form of *a priori* information. On the other hand, the robot would suffer from cumulative drift errors, because such errors could never be compensated by proprioception alone. The argumentation given so far and the experimental validation in Sect. 5 provide strong evidence that Duckett's statement as is no longer holds. The potential objection that knowing the route graph in advance should be classified as non-proprioception eventually will disappear when ROUTELOC is extended by a proprioceptive exploration component for map building, discussed in section 5.4.3.

Thus, the drawback of only relying on proprioceptive data is *not* that it hinders self-localization. The problem to be tackled is that the available amount of information received from the odometry is not very rich, which fosters the problem of *perceptual aliasing* (cf. Sect. 4.3.4).

4.2.2   Environment Model

The reference used for localization is the so-called *environment model*. An environment model is a description of the task-relevant characteristics of the world the robot operates in. Examples from daily life are city maps, hiking descriptions, or subway maps. In order to find a way to a goal point, it is necessary to locate one's current real world position in the model and subsequently plan a way to the goal point.

To make this localization possible, the model used for describing the current situation of the robot must be "compatible" with the model used for the description of the robot's environment. As a negative example, consider a scenario where your situation is defined by the visual perception of features in the environment, e. g. street signs in Moscow. To determine your position on a city map, you have to match the Cyrillic letters, e. g., on the street sign "Кразная Площадь", with the international transcription in Latin letters ("Kraznaya Ploshchadj") used in some city maps, or, even worse, with the translation ("Red Square") used in others. This is rather difficult if you do not know how to pronounce or translate Russian words. In this example, the matching between the situation model (visual perception of a street sign with Cyrillic letters) and the environment model (city map annotated with street names in international transcription or English translation) has failed. The lesson learned from this example is that the environment model used for self-localizing the robot has to be compatible with the generalized route descriptions presented in Sect. 4.2.1.

The second requirement on the environment model (which, by the way, also applies to the situation model) is that it should be appropriate for the intended application scenario of the robot. Developing service robots, especially rehabilitation robots, usually means the development of low cost devices. Therefore, the equipment used for self-localization should be as sparse as possible. Nevertheless, mobile service robots such as cleaning or surveillance robots, and smart wheelchairs often have to cover a large operation space. This means that the self-localization approach must work in large-scale environments such as complex buildings, university campuses or hospital areas. Especially in the context of rehabilitation robots the environment cannot easily be changed, e. g., by mounting artificial landmarks or beacons at decision points, because they are often part of public buildings. Furthermore, environment changes are very expensive. As a consequence, an approach is needed that requires only minimal sensor equipment, works in unchanged environments, and is able to operate reliably in large-scale environments.

Nevertheless, there are recent research activities that pursue a completely different approach by adding small active beacons to the environment that transmit their ID and allow the passing robot to use them as landmarks (Kantor and Singh, 2002). Since the system does not pre-require that the robot's transponder and the active beacons are in line of sight, it works robustly even in cluttered environments. However, the effort to equip a large-scale environment with these beacons — and determine their positions — is not neglectable.

Taking into account these aspects, a topological map that is enhanced with certain metric information appears as an adequate representation of the environment in this context.

The Route Graph

The previous sections suggest the modeling of the environment as a hybrid topological-metric graph structure. The so-called *route graph*, which serves as RouteLoc's environment model, is such a (non-directed) graph structure.

In the following, the nodes of a route graph correspond to decision points in the real world: corridor corners, junctions or crossings. The edges of a route graph represent straight corridors that connect the decision points. In addition to the topological information, the route graph contains (geo-)metric data about the length of the corridors and about the rotation angles between the corridors. As an example, consider the route graph depicted in Fig. 4.3. The sketch in Fig. 4.3a shows the second floor of the MZH building of the Universität Bremen. In Fig. 4.3b, the corresponding route graph is depicted. The nodes of the graph represent decision points in the world, the edges correspond to straight corridors. The route graph consists of 22 nodes (decision points) and 25 edges (corridors) connecting them.

The basic form of the route graph used here is complex enough to serve as environment model for large-scale self-localization. However, it is only a small fraction of the powerful concept that Werner et al. (2000) called "route graph" as well. They were the first to use the notion *route graph* not only in the meaning of "network of routes", but constructed a whole theory on modeling navigational knowledge around a route graph. Their route graph is defined as a hierarchical data structure that subsumes the concepts of *routes* and *places*. Following the terminology used by Werner et al. (2000), a route is a sequence of directed *route segments*. Each route segment connects two *places*. Each place is a "tactical decision point" and directly corresponds to a decision point as used in this thesis. Both, places and route segments may be annotated with attributes. Werner et al. (2000) define a *route graph* to be a directed graph with a set of places (nodes) and a set of route segments (directed edges). Such a route graph may comprise a whole hierarchy of routes. For instance, a distinct real world location such as the Münster main station may be represented by different places, depending on which kind of route is referred to: the train line to Bremen, the bus line 12 to the Tannenhof, or the pedestrian walk to the Prinzipalmarkt in the city center. As is pointed out in Sect. 5.4.3, it would be interesting to extend RouteLoc to such a multi-layer form of route graph.

Interestingly, the notion "route graph" is also common in other research areas. For instance, the file format specification for the Virtual Reality Modeling Language VRML describes a "route graph" that is a data-flow network of event communication (VRML97). The route graph specifies behavior in VRML.

The representation of the environment as a simple route graph is formally very similar to Voronoï diagrams that are often used in mobile robot navigation, see the discussion in Sect. 3.2. The advantage of the route graph is that it is defined independently of the current obstacle situation. Thus, there is no need to rely on input from proximity sensors as is necessary for the Voronoï diagram based approaches (a Voronoï diagram is defined on the basis of the sensor-perceived distance of the robot to objects in its environment).

Since the route graph (environment model) has to be matched with route generalizations (situation model), it is advantageous *not* to implement the graph as a set of nodes that are connected by the edges, but as a derivable structure, the set of so-called *junctions*:

Definition 4.1 (Junction) *Given a planar graph N=(V,E), a junction j is a 5-tuple*

$$j := (H, T, \gamma, d, I)$$

a)   b)

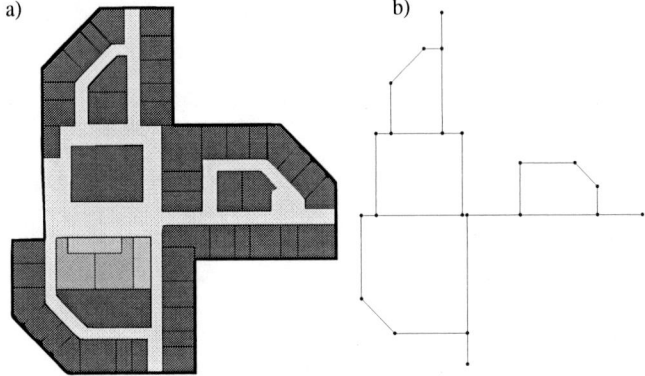

Figure 4.3: *Route Graph Example.*

In Def. 4.1, $H \in V$ ("home" of the junction j) and $T \in V$ ("target" of j) are graph nodes that are connected by a straight corridor of length d, as shown in Fig. 4.4. The set I consists of all incoming junctions j_i' that lead to j, i.e. $I = \{(H', H, \gamma', d', I')\}$. The function $in(j)$ selects the incoming junctions of j, i.e. $in(j) = I$. The signed angle γ is the rotation angle between the prolongation of an outgoing segment of some j_i' to the outgoing segment of junction j, i.e. it denotes by how many degrees a robot has to turn in order to travel through j. For left turns, γ is positive; for right turns, γ is negative; $\gamma = 0$ means that j is a so-called "straight junction", e.g. the T-bar of a T-junction (cf. Fig. 4.4). The symbols used for annotating the figure correspond to those used in definition 4.1. The underlying route graph is depicted by thin grey lines. The outgoing (directed) segment of the junction is shown as a solid black arrow. Note that outgoing segments of junctions are directed, i.e. junctions are one-way connections between route graph nodes. As will be shown in Sect. 4.3.2, the corners of a route generalization are compatible with the junctions of the route graph in that they can be matched and assigned with a real number representing a similarity measure.

Please also pay attention to a possible future enhancement of the approach that is discussed in Sect. 5.4.3: each junction can be associated with a set of so-called feature triples (f, q, p). Such a feature triple indicates that the robot will observe feature f (e.g., a landmark) with a probability of p after traveling a distance of q cm in the outgoing segment of the junction.

Based on Def. 4.1, a route graph G in its "junction representation" is defined as the set of all junctions:

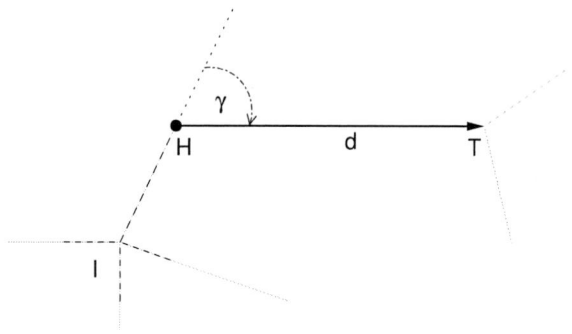

Figure 4.4: *Junction in a Part of the Route Graph.*

Definition 4.2 (Route Graph in its Junction Representation) *A route graph G in its junction representation is a set of all junctions that are connected:*

$$G = \{ j = (H, T, \gamma, d, I) \mid \exists j' \in G : j' \neq j \wedge j' \in in(j) \}$$

For the ease of reading, the suffix "in its junction representation" is omitted for the rest of this thesis when referring to a route graph, provided that the meaning is clear because of the context. As a rule of thumb, the notion "route graph" normally refers to the junction representation as defined in Def. 4.2. However, most of the following figures show route graphs as nodes connected by edges since depicting a route graph in its junction representation is not that easy.

In contrast to grid-based representations, such a data structure is much easier to handle with respect to the required amount of computing time and memory. For example, the campus environment used for experiments in the results section (see Ch. 5 and Fig. 5.1 on page 98) is coded as a list of only 144 junctions (see Fig. 5.4).

Even though the route graph is originally intended to model networks of straight corridors, a lot of everyday situation configurations can be represented. Figure 4.5 shows how three different real world scenarios could be represented by a route graph. Figure 4.5a represents the ideal case where the route graph on the right of the sub-figure a) is a perfect representation of the corridor-dominated environment sketched on the left. Figure 4.5b shows a more difficult case, because larger places are more problematic to be presented by a route graph. It helps to represent the place as a set of completely connected decision points. Figure 4.5c shows that curves may be modeled as polygons. As will be shown in Sect. 4.4.6, the route graph's complexity does not depend on its spatial extent but on the number of represented decision points.

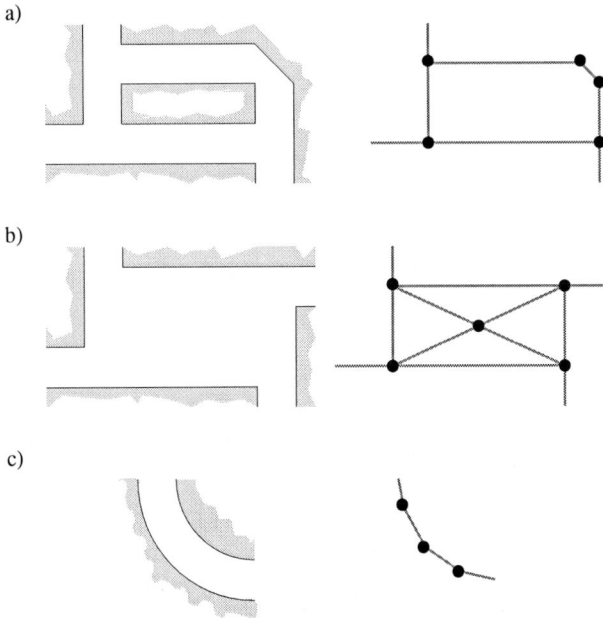

Figure 4.5: *Modeling the Environment as Route Graphs — Three Examples.*

4.3 ROUTELOC: an Overview

This section is meant to explain how generalized route descriptions as situation model and a route graph as environment model are used for the absolute self-localization of a robot in large-scale environments. A sketch of ROUTELOC is presented to explain the basics of the algorithm. The simplifying assumptions made in this section for clarity are subsequently dropped in the detailed description of ROUTELOC in Sect. 4.4.

4.3.1 ROUTELOC in a Nutshell

The basic idea of the self-localization approach is to match an incremental generalization of the currently traveled route with a route graph. This matching process provides a hypothesis about the robot's current position in its environment.

With a probabilistic approach RouteLoc continuously determines the corridor (represented by an edge in the route graph) in which the robot is most likely located at that very moment in time. Since the distance already traveled in this "candidate corridor" is also known, an additional metric offset can be estimated. As a result, the position of the robot within the corridor is found precisely enough for most global navigation tasks. The precision is limited by about half of the width of the corridor the robot is located in (see Sect. 4.3.4).

The route-localization algorithm described here has to know the route graph in advance. Nevertheless, Sect. 5.4.3 discusses how future work is going to deal with exploring the robot's environment from scratch and building a route graph on the fly, thus dealing with the SLAM-problem briefly introduced in Ch. 3.2 on page 33.

4.3.2   Matching Route and Route Graph

Because of the dualism between a junction in the route graph and a corner in the generalized route, the chosen situation model and the environment model are compatible. Thus, self-localizing a robot by matching a generalized route with a route graph should in principle be straightforward. Nevertheless, there are some pitfalls that have to be paid attention to.

Since RouteLoc has to deal with real data, there are almost no "perfect matches". That means, even if the robot turned by exactly 90° at a crossing, the angle of this corner as calculated by the route generalization will almost certainly differ from 90°. This is mainly due to odometry errors. On the other hand, two corridors that meet in a perfect right angle in the route graph may well include an angle of only 89.75° in reality. These uncertainties have to be coped with adequately, as discussed in the following subsections.

A second topic worth considering is the complexity of the matching process: At least in theory, a route can consist of arbitrarily many corners. Therefore, matching the whole generalized route with the route graph in each computation step is not feasible, because—at least in theory—this would require an arbitrarily long period of computing time. A solution to this problem is presented in the following subsections.

Within this section, it is assumed that every corner existing in reality is detected by the generalization algorithm *and* that every corner detected by the generalization algorithm is existing in reality. As mentioned earlier, this assumption is simplistic and unrealistic. Nevertheless, it is reasonable here in order to simplify the explanation of the basic structure of RouteLoc. The details of RouteLoc are thoroughly discussed in Sect. 4.4.

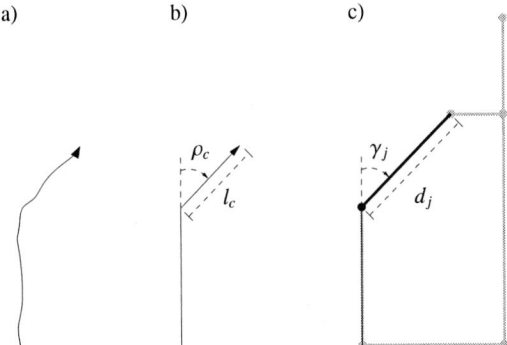

Figure 4.6: *Direct Match of Route Corner and Route Graph Junction.*

Direct Match of Route Corner and Graph Junction

Determining whether a whole route matches a subgraph of the route graph is an intractable problem, as is shown below. It is much easier to match only a single corner with a route graph junction: A *direct match* of a route corner c and some junction j in the route graph is possible (cf. Fig. 4.6). The figure shows the trajectory traveled by the robot as recorded by its odometry (a.); the corresponding route generalization (b.), where the final corner is defined by the rotation angle from the previous to the current segment and the route segment's length; and a part of a route graph with a highlighted junction that is to be directly matched with the route corner (c.).

As mentioned above, a binary decision of whether or not c and j match is not adequate in this situation. Thus, a probabilistic similarity measure is introduced that describes the degree of similarity between the route corner and the junction as a real number between 0 and 1.

The similarity measure m_d for the direct match of a route corner with a route graph junction is defined as

$$m_d(c, j) = s_l(c, j) \cdot s_\alpha(c, j) \qquad (4.2)$$

In (4.2), the similarity s_l of the length of the outgoing segment of j and the route segment of c, is defined as $s_l(c, j)$ with

$$s_l(c, j) = sim\left(\frac{|l_c - d_j|}{d_j}, 0\right) \qquad (4.3)$$

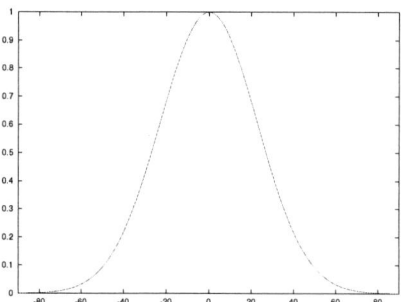

Figure 4.7: *Similarity Measure for the Rotation Angle.* The graph shows the similarity measure for the angles (y-axis) as a function of the route angle for a junction angle of $0°$.

In (4.3), l_c is the length of the outgoing route segment of c; d_j is the length of the outgoing corridor of junction j (see Fig. 4.6). The longer the corridor, the larger the deviation may be for a constant similarity measure.

The similarity of the corresponding rotation angles, $s_\alpha(c, j)$, is likewise defined as

$$s_\alpha(c, j) = sim\left(\|\gamma_j - \rho_c\|, 0\right) \qquad (4.4)$$

In (4.4), γ_j is the rotation angle between the two segments of junction j, and ρ_c is the rotation angle of the final route corner c (see Fig. 4.6). Note that the result of this subtraction is always shifted into the interval $[0, \ldots, \pi]$, as indicated by the $\|\ldots\|$ notation. Please note also that these equations will be refined in the following sections in order to cover some special cases that will be introduced below.

In (4.3) and (4.4), the function $sim(x, m)$ returns a value of the interval $]0 \ldots 1]$, which describes the similarity of x and m. The function sim is a normalized Gaussian function with mean m and standard deviation σ. It produces high similarity values, when the lengths or angles, respectively, differ only little. It produces low similarities, when the lengths or angles, respectively, are completely different.

In the current implementation that has been used for the experiments presented in the results chapter Ch. 5, sim is defined as

$$sim(x, m) = e^{\pi(x-m)^2} \qquad (4.5)$$

This is a Gaussian function with σ set to $1/\sqrt{2\pi}$ such that the vertex of the curve is always at $(m/1)$, i. e. $sim(m, m) = 1$. Since the relations are about the same, the same function sim can be used for the similarity measures of both, the segments' lengths and the rotation angles.

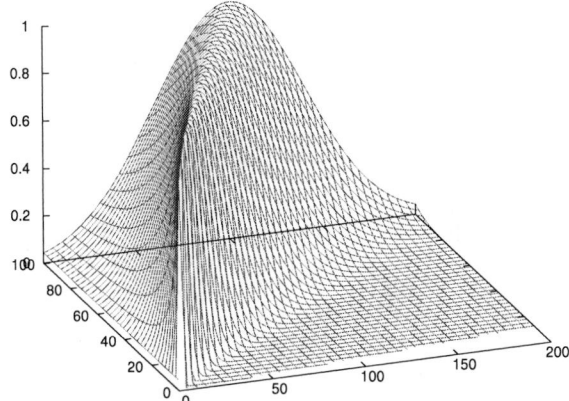

Figure 4.8: *Similarity Measure for the Segment Lengths.* The 3D-figure depicts the similarity measure for the segments' lengths (z-axis) as a function of the length of the junction's corridor and of the current route segment.

As a result, the similarity measure for the rotation angles behaves as depicted in Fig. 4.7. The figure shows a case where the junction angle is $0°$, i.e. it is a straight junction. The plotted function marks the similarity values for all possible rotation angles of the route corner between $90°$ to the left ($+90°$) and $90°$ to the right ($-90°$). Please note that larger deviations yield a result very close to zero. The smaller the difference between the corner angle and the junction angle, the closer the similarity is to 1.

Figure 4.8 shows the similarity measure for the segments' lengths. Here, the horizontal axis represents the length already traveled in the final route segment (possible values 0 m – 200 m). The third axis (domain 0 m - 100 m) represents the length of the junction. This means, in the depicted situation, that the robot may overshoot by 100%.

To illustrate the effect of the similarity measures, consider Tab. 4.1. As examples, seven cases are discussed here:

1. This case shows a perfect match. Both, rotation angles and segments' lengths fit perfectly. Therefore, the similarity value is 1.0.

2. This case shows a perfect match with respect to the angles. The distance traveled in the current route segment is only 9 m, whereas the junction's segment is 10 m in length. Thus, the traveled distance falls short by 10%, which results in a still relatively good similarity measure of 0.96. Please

Table 4.1: *Similarity Measure in Examples.* The similarity measure $m_d(c, j)$ depends on the rotation angles γ of junction j and ρ of corner c on the one hand, and on the length d_j of the outgoing junction segment and l_c, the length of the route segment of corner c on the other hand.

No.	Rotation Angle			Segment's Length			Similarity
	γ	ρ	s_α	d_j	l_c	s_l	$m_d(c, j)$
1.	90°	90°	1.0	1000	1000	1.0	1.0
2.	90°	90°	1.0	1000	900	0.962	0.962
3.	90°	80°	0.909	1000	1000	1.0	0.909
4.	90°	90°	1.0	1000	250	$5.26{\cdot}10^{-13}$	$5.26{\cdot}10^{-13}$
5.	90°	22°	0.012	1000	1000	1.0	0.012
6.	90°	80°	0.909	1000	900	0.962	0.874
7.	90°	22°	0.012	1000	250	$5.26{\cdot}10^{-13}$	$6.29{\cdot}10^{-15}$

note that overshooting the junction segment's length is symmetric.

3. This case shows a perfect match in the segments length component. The angles differ only slightly by 10°. As a result, the similarity measure drops to about 90%.

4. This case again shows a perfect match with respect to the angles. But this time, the traveled distance is only 25% of the junction's segment length. As a result, the similarity measure is almost zero.

5. This case shows a perfect match in the length component, but a significant deviation with respect to the angles. As a result, the similarity measure is rather low.

6. This case shows slight deviations in both components. Nevertheless, the resulting similarity measure is still relatively high.

7. This case shows that significant deviations in both components result in a similarity measure of almost zero.

Please note that the exact definition of the similarity measures does not influence the performance of the self-localization algorithm. "Tuning" the parameters may have some beneficial effect in a specific situation, but may result in a worse behavior in another situation. The important point is that slight deviations are tolerated whereas significant deviations in one component cause the similarity measure to drop significantly.

Matching the Whole Route

After having defined the direct matching for a single corner, the similarity measure has to be extended to complete routes. When a route $R = \langle c_0, \ldots, c_n \rangle$ with $n > 1$ is to be matched with a junction j, it has to be detected, whether there is a direct match between corner c_n and junction j, *and* whether there is one between c_{n-1} and some j' with $j' \in \text{in}(j)$, *and* whether there is one between c_{n-2} and some j'' with $j'' \in \text{in}(j')$, *and* so on. If such a sequence of junctions of the route graph can be found, the whole route R can be matched.

Thus, the *matching quality* of a complete route R with respect to a specific route graph junction j is defined as follows:

Definition 4.3 (Matching Quality) *Given a route* $R = \langle c_0, \ldots, c_n \rangle$ *with* $n > 0$ *and a junction* j *of the route graph* G *(* $j \in G$ *), the matching quality* $m(R, j)$ *of* R *with respect to* j *is defined as*

$$m(R, j) = \max \left\{ \prod_{i=0}^{n} m_d(c_i, j_i) \; \middle| \; \exists \langle j_0, \ldots, j_n \rangle : j = j_n \wedge j_{k-1} \in \text{in}(j_k) \right\} \quad (4.6)$$

and, for $n = 0$, $m(R, j)$ *is defined as*

$$m(R, j) = m(\langle c_0 \rangle, j) := m_d(c_0, j) \quad (4.7)$$

Definition 4.3 states that every possible sequence of length $n + 1$ of route graph junctions is considered that fulfills two requirements: the final junction of the sequence must be j and the sequence must be "traversable", i. e. the k-th junction in the sequence must be an incoming junction of the $(k + 1)$-st junction of the sequence. Since such a sequence consists of as many junctions as there are route corners, the matching quality with respect to a certain sequence can be determined by calculating the product of the direct matching qualities of the sequence junctions and the corresponding route corners. The overall matching quality of the route R and junction j is the maximum of all these products.

The number of such sequences grows exponentially with the length of the route and with the number of junctions in the route graph. Therefore, the defining equation (4.6) is inadequate for a real-time localization approach. Furthermore, it is no solution to simplify the task by only considering the last, say, three corners of a route for the matching process, since the decisive difference may have occurred at the beginning of the robot's journey: Figure 4.9 shows such a situation. In Fig. 4.9a, a route graph is depicted. The highlighted junction is to be matched with two different route generalizations shown in Figs. 4.9b and c. While the final corners of both perfectly fit the route graph, their first real corner makes the difference: The route depicted in Fig. 4.9b does not match the route graph whereas the one in Fig. 4.9c does.

a)   b)   c)

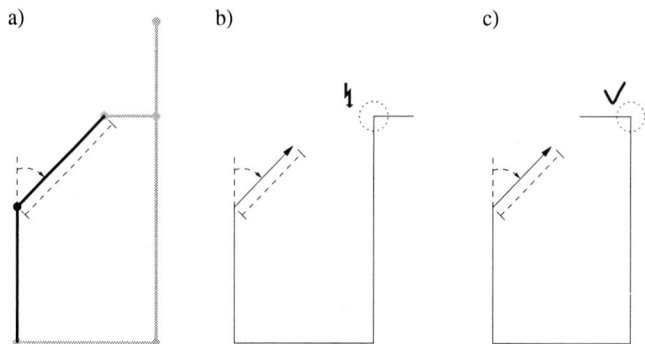

Figure 4.9: *"Induction Step": Matching of the Rest of the Route.* A match or a mismatch, respectively, in the early phase of the route can decide about the current matching quality, as depicted here: The left figure shows the route graph with a highlighted junction in the upper left corner. The middle and the right figure depict two route generalizations that differ only in their first real corner (marked by a dashed-circle).

Fortunately, there is a better solution that significantly reduces the complexity of calculating the matching quality: following the idea of the incremental route generalization, the matching quality can be defined inductively. In order to determine $m(\langle c_0, \ldots, c_n \rangle, j)$, it is sufficient to know $m_d(c_n, j)$ and $m(\langle c_0, \ldots, c_{n-1} \rangle, j')$ for all $j' \in in(j)$. As a result, the defining equation (4.6) can be refined to

$$
\begin{aligned}
m(R, j) &= m(\langle c_0, \ldots, c_n \rangle, j) \\
&= \max \left\{ \prod_{i=0}^{n} m_d(c_i, j_i) \;\middle|\; \exists \langle j_0, \ldots, j_n \rangle : j_n = j \wedge j_{k-1} \in in(j_k) \right\} \\
&= m_d(c_n, j) \cdot \max \left\{ \prod_{i=0}^{n-1} m_d(c_i, j_i) \;\middle|\; \exists \langle j_0, \ldots, j_{n-1} \rangle : \right. \\
&\qquad\qquad\qquad\qquad\qquad\qquad \left. j_{n-1} \in in(j) \wedge j_{k-1} \in in(j_k) \right\} \\
&= m_d(c_n, j) \cdot \max_{j' \in in(j)} \left\{ \max \left\{ \prod_{i=0}^{n-1} m_d(c_i, j_i) \;\middle|\; \exists \langle j_0, \ldots, j_{n-1} \rangle : \right. \right. \\
&\qquad\qquad\qquad\qquad\qquad\qquad \left. \left. j_{n-1} = j' \wedge j_{k-1} \in in(j_k) \right\} \right\} \\
&= m_d(c_n, j) \cdot \max_{j' \in in(j)} \left\{ m(\langle c_0, \ldots, c_{n-1} \rangle, j') \right\} \qquad (4.8)
\end{aligned}
$$

Thus, the matching quality $m(R, j) = m(\langle c_0, \ldots, c_n \rangle, j)$ is defined as

$$m(\langle c_0, \ldots, c_n \rangle, j) = \begin{cases} m_d(c_0, j) & , n = 0 \\ m_d(c_n, j) \cdot \max_{j' \in in(j)} \{m(\langle c_0, \ldots, c_{n-1} \rangle, j')\} & , n > 0 \end{cases} \quad (4.9)$$

Calculating this recursion in every step is still impractical because the computational complexity would depend on the length of the route. Fortunately, the recursive function call in (4.9) can be avoided, if each junction is assigned with the probability value for having been in one of its incoming junctions before. According to (4.9), this "history factor" is called the *incoming matching quality* $m_{in}(j)$ which is defined as

$$m_{in}(j) = \max_{j' \in in(j)} \{m(\langle c_0, \ldots, c_{n-1} \rangle, j')\} \quad (4.10)$$

Section 4.3.3 explains how $m_{in}(j)$ is maintained and how the recursive function call is avoided.

By applying Def. 4.3 to the current route and every route graph junction, the junctions are assigned with a matching quality. The maximum of all the matching qualities provides a hypothesis which junction most likely hosts the robot. This junction is called *candidate junction* j_c for a route R.

$$j_c(R) = \text{argmax}_{j \in G}\{m(R, j)\} \quad (4.11)$$

Figure 4.10 presents a step-by-step visualization of the matching process: In the initial situation, no information about the robot's potential location is available. Therefore, every junction in the graph can host the robot with the same likelihood. This is indicated by the edges underlined in grey in the route graph that is shown in the upper row of the figure. After the robot traveled some distance in a corridor (cf. Fig. 4.10b), three edges in the graph are identified in which the robot cannot be located. The route segment just traveled is longer than the corridors represented by these edges. After completing the first turn (90° to the left, see Fig. 4.10c), basically only three possibilities remain: Either the robot started in a corridor that is represented by one of the two facing edges depicted vertically in the lower part of the route graph or it started horizontally and its location is in the upper part of the graph afterwards. In Fig. 4.10d, another left turn yields no new information and thus no reduction of the possible robot locations. As shown in Fig. 4.10e, the situation clarifies after the following turn: the location of the robot is determined by figuring out a unique candidate junction.

4.3.3 Propagation

The previous section motivated that the incoming matching quality $m_{in}(j)$ defined in (4.10) cannot be re-calculated in every update step but has to be handled more

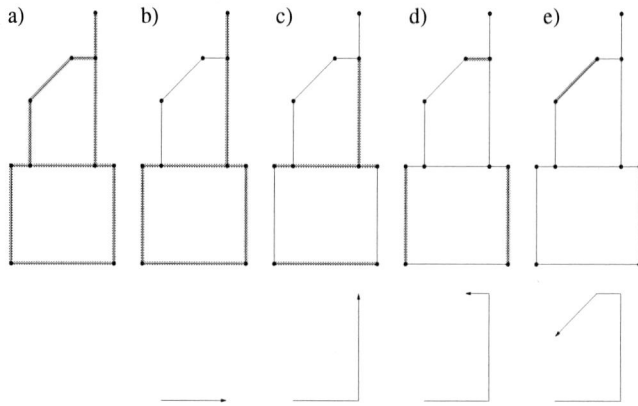

Figure 4.10: *Matching of Route Generalization and Route Graph.*

efficiently instead. In order to obtain a closed expression for $m(R, j)$ without recursive function calls, $m_{in}(j)$ has to be associated with each junction and updated whenever necessary. Note that m_{in} remains unchanged as long as no new route corner is detected by the route generalization algorithm. But with each detected route corner, the semantics of $m_{in}(j)$ changes. This is because it refers to a different part of the route after such a detection. That means that after the detection of a new route corner the probability of being in the outgoing segment of a junction k becomes the probability of having been in the incoming segment of some junction k' with $k \in in(k')$ before.

The process of transferring the matching qualities between the junctions in case of a corner detection is called *propagation*. Please note that the propagation process is independent of the similarity measures defined in (4.3) and (4.4) on page 55. It is a generic approach to update a probability distribution over a set of junctions which represent a route graph.

The propagation process is illustrated in Fig. 4.11. The figure shows four snapshots a)-d) of a route traveled by the robot in a triangular environment. The left part of each row shows the generalized trajectory as recorded by the robot. The arrow indicates the current position. The right part of each snapshot depicts a route graph that consists of six junctions. Each junction is assigned with two probability values, depicted as partly filled columns. The left column (filled with dark grey) indicates the direct matching quality of the final route corner with this junction. The right column (filled with light grey) describes the probability of having been in the incoming segment of this junction before (the incoming matching

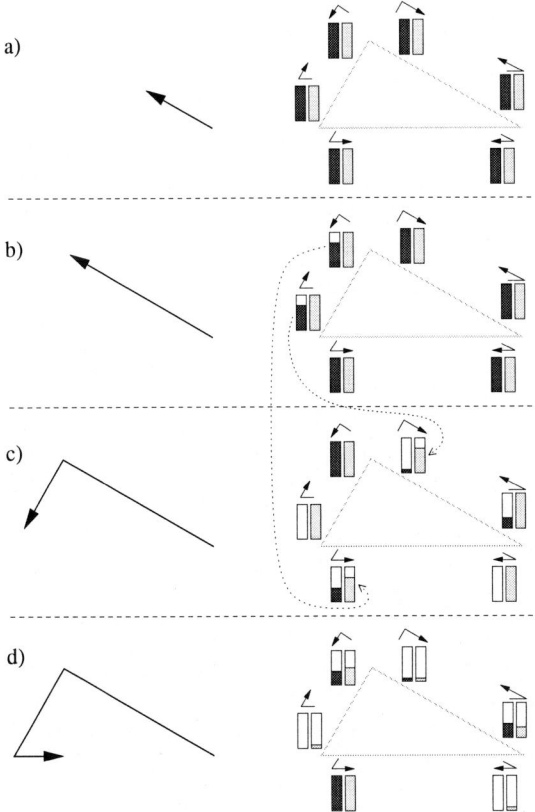

Figure 4.11: *Propagation of Probabilities.*

quality). A completely filled column stands for a 100% match, an empty column means something below 10% (but more than 0%). The arrows above the probability columns indicate the junction, e. g. the columns in the lower left corner of the route graph belong to the junction that leads from the left corridor to the lower corridor with a rotation angle of about 120° to the left. The dotted arrows between row b) and c) indicate the propagation. Please note that the figure is not based on real data and the depicted probability values are not normalized, i. e. the probability values do not sum up to 1.0 and should only be interpreted as a qualitative measure.

In the situation depicted in Fig. 4.11a, the robot has already driven some straight path. Since the traveled distance is shorter than any junction segment in the route graph, the probabilities are uniformly distributed over all junctions. After some more meters, the situation depicted in Fig. 4.11b arises: the straight segment traveled by the robot is already longer than the outgoing segment of the two junctions covering the left corridor. Therefore, their direct matching qualities drop below 1.0. The other junctions still give a 100% match. After detecting the first real corner in the route, the situation of Fig. 4.11c occurs: The matching qualities are propagated to the adjacent junctions. Note that the most likely junction is already correctly identified, while some other junctions are still rather probable. The detection of the second corner clarifies the situation, as depicted in Fig. 4.11d. After another successful propagation, the most likely junction is correctly determined to be the one corresponding to the lower corridor (directed to the right) of the graph.

The example shown in Fig. 4.11 only visualizes the propagation of the probability values through the route graph. Since each junction in the depicted route graph comprises only one incoming junction, it is prescribed the probability value of which junction has to be propagated. If, instead, there is more than one incoming junction for a junction j, a decision has to be made, which "history" has to be believed, i. e. which incoming matching quality has to be propagated. The detailed process is visualized in Fig 4.12.

The figure depicts the propagation process in five rows a) to e). From left to right, each row shows the route generalized so far, a part of the route graph with some junctions labeled by small arrows, and a table with probability values of these junctions. As an example, consider row a): the final corner of the route (marked by a grey shadow) is to be matched with the three junctions depicted in the route graph. Junction j_1 (marked by the 90° arrow) is the best match for the corner, both, with respect to the direct matching quality m_d and to the incoming matching quality m_{in}. Please note that $m_d(j_1) = 1.0$, because the rotation angles of the corner and the junction perfectly fit, i. e. $s_\alpha(c_n, j_1) = 1.0$. The length of the outgoing route segment is shorter than the junction's segment, but this is tolerated since it is the final route segment which is not yet fixed in length (cf. Sect. 4.2.1), thus also $s_l(c_n, j_1) = 1.0$. In Fig. 4.12b, the direct matching quality for j_1 drops below 1.0, because the route segment is longer than the corresponding junction segment. Then, the route generalization algorithm detects a new corner c_{n+1} in the route (see Fig. 4.12c). As a consequence, the propagation process is initiated. It consists of three steps:

Matching the penultimate corner. The now penultimate corner c_n is fixed in length and rotation angle. Therefore, this corner is matched with the junc-

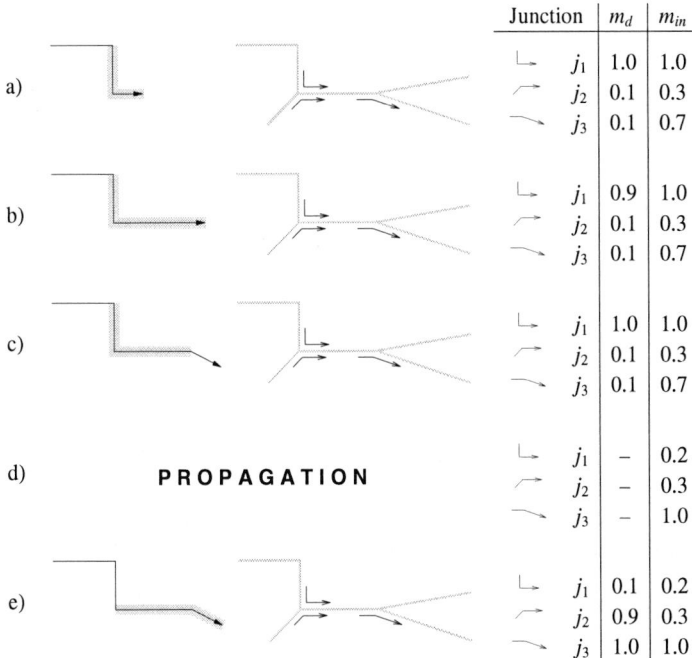

Junction		m_d	m_{in}
a)	j_1	1.0	1.0
	j_2	0.1	0.3
	j_3	0.1	0.7
b)	j_1	0.9	1.0
	j_2	0.1	0.3
	j_3	0.1	0.7
c)	j_1	1.0	1.0
	j_2	0.1	0.3
	j_3	0.1	0.7
d) PROPAGATION	j_1	–	0.2
	j_2	–	0.3
	j_3	–	1.0
e)	j_1	0.1	0.2
	j_2	0.9	0.3
	j_3	1.0	1.0

Figure 4.12: *Determining $m_{in}(j)$*

tions in the route graph (see the highlighted corner in Fig. 4.12c). Note that $m_d(j_1)$ is 1.0 again, because the corner in the route is located at a place where the segment's length gives a perfect match. Please also note that m_{in} is still unchanged for all junctions.

Propagation. The propagation process checks for every junction k, which incoming junction k' provides the most probable path to k. The matching quality of this junction k' is subsequently used as incoming matching quality of k, i. e. $m_{in}(k) = \max_{k' \in in(k)} m(\langle c_0, \ldots, c_n \rangle, k')$. As a consequence, $m_{in}(j_3)$ is set to $\max\{m(R, j_1), m(R, j_2)\} = 1.0$ (see Fig. 4.12d). Note that this step is a simple process which does not require any recursive calculation. It is merely necessary to determine the maximum of a set of real values.

Matching the new current corner. In the final propagation step, the new current corner c_{n+1} has to be matched with the updated junctions. In Fig. 4.12e, j_3 is the candidate junction, because the slight right turn in the route gives a

a)   b)

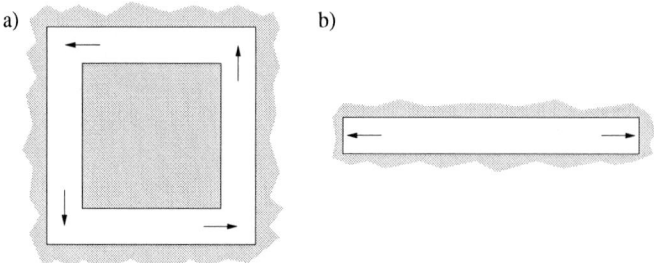

Figure 4.13: *Matching Problems in Ambiguous Environments.*

perfect direct match $m_d(c_{n+1}, j_3)$ and $m_{in}(j_3)$ has been set to 1.0 during the propagation step.

After completing these steps, the probability values of all junctions are normalized.

Note that the propagation process includes a Markov step in as much as the propagation reduces the information available about the initial phase of the route traveled so far to the fact that it is so and so likely to having been in the incoming junction of the considered junction before. Nevertheless, this Markov property is violated by the extensions that have to be added to ROUTELOC as shown in Sec. 4.4.

4.3.4   Estimating the Robot's Position

As shown so far, the matching process assigns a probability value to each junction. This value, the matching quality of the route traveled so far with respect to the specific junction j, describes how well the route can be embedded in a part of the route graph that ends in j. The junction with the highest probability value is the one the robot is most likely located in. As defined in (4.11), this junction is called candidate junction. Since the distance traveled in the final route segment can be used to calculate an offset that has already been traveled in the candidate junction's outgoing segment, ROUTELOC is able to estimate a metric position of the form "The position is in the corridor that leads from decision point **A** to decision point **B**, x meters past **A**". Due to the modeling of the environment as a route graph, the position estimate within a corridor is one-dimensional, i.e. the lateral deviation is not considered.

One could argue that this metric information is superfluous for the user or for higher level navigation modules, because the corridors *between* the decision points are by nature free from decisions such as turning to a neighboring corridor. Thus, no detailed information about the robot's location should be required. For

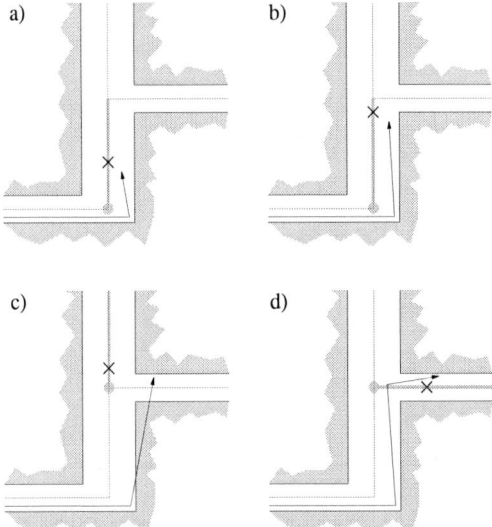

Figure 4.14: *"Generalization Delay" When Turning From one Corridor to Another.*

instance, Nourbakhsh et al. (1995) follow this idea. They come to the conclusion, that only having the topological information about the current corridor is often not enough. Therefore, they propose to also model the metric position within the corridors.

The same applies here, where the metric information is indispensable for two reasons: First, not every location that is important for the robot's task can be modeled as a decision point. Consider, e. g., some cupboard a wheelchair driver has to approach in a corridor. Second, when traveling autonomously, the robot often has to start actions or local maneuvers in time, i. e. they have to be initiated at a certain place in the corridor, maybe well before the relevant decision point can be perceived by the robot. This would be impossible without the metric information. Another point is that the use of the already mentioned features will rely on this metric information. This is described in Sect. 5.4.3.

The rest of this section discusses some aspects that are relevant for a successful position estimate.

Ambiguous Environment. In some situations, the structure of the environment could turn out to be inadequate to this route-localization approach in its current version. For instance, in a square environment such as the one depicted in Fig. 4.13a, RouteLoc will fail, because every junction remains equally likely to host the robot even if the robot moves through the corridors for a long time. This is equivalent to the problem of *perceptual aliasing*. This notion was coined by Whitehead and Ballard (1991), referring to situations that look identical to the robot but are in fact different and often also require different actions. When traveling in the square environment, four position estimates that are equally likely would be favored, no decision for a specific corridor would be possible. The four arrows depicted in Fig. 4.13a indicate an example for four such ambiguous positions.

Similarly, in the straight corridor in Fig. 4.13b, the algorithm is almost lost, because it has no means to infer where the robot started. Nevertheless, the longer the robot moves along the corridor, the less estimates are valid, simply due to the length of the trajectory already traveled. But even, if the robot traveled a straight route segment of about the corridor's length, RouteLoc still would generate two hypotheses about the robot's position, one at each end of the corridor. The arrows in the figure visualize an example for two such ambiguous positions.

Note that the problem of perceptual aliasing is not unique to RouteLoc. It is a general problem that is more obvious if the robot only relies on a small amount of sensor input. Due to its minimal sensor usage, RouteLoc is prone to perceptual aliasing.

"Generalization Delay" when Changing Corridors.   Due to the nature of the generalization algorithm, there exists a certain delay before the change of corridors can be detected. To illustrate this, Fig. 4.14 shows a sequence of four situations of a travel where the robot first performs a left turn, and then—at a T-junction—it turns right. This is depicted by the solid black arrow which shows the generalized route driven by the robot, i. e. the arrow indicates the position of the robot in the environment. The route graph, which is also embedded in the environment, is shown as a thin dark grey line. The candidate junction (the most probable in the route graph) is depicted as a thick light grey line for its outgoing segment and a circle for its corresponding route graph node. The estimated position of the robot is marked by a cross indicating a certain offset in the candidate junction of the route graph.

In Fig. 4.14a, the generalization of the traveled route is correctly matched with the route graph. The estimated position differs only slightly from the real position (cf. the paragraph on precision below). In Fig. 4.14b, the robot almost reached the T-junction. The localization is still correct. In Fig. 4.14c, the robot already changed corridors by taking the junction to the right. But the generalization algo-rithm has not yet been able to detect this, because it still can construct an "accep-tance area" for the current robot position within the same corridor as before (see

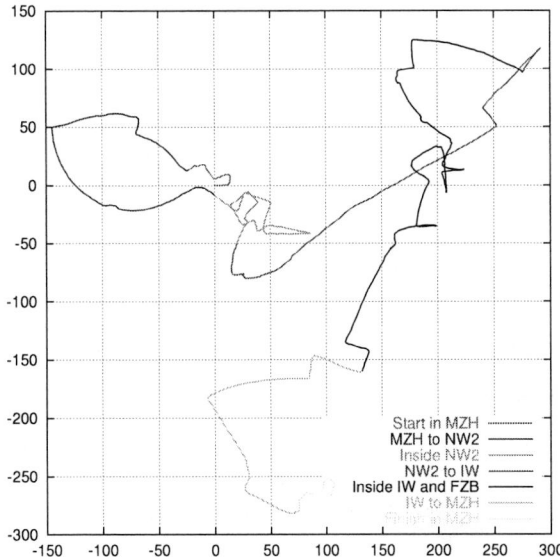

Figure 4.15: *Log Data of the Wheelchair's Odometry.* Axes' unit is meter. The route starts at (0/0). Tracks in different buildings are depicted by different line types. Even though start and end point are identical in reality, they differ by 290 m in the recorded data. The key labels refer to the building names. The order of the route parts is chronological, i.e. the robot starts in the MZH building on route part MZH (solid line) and also finishes there (short dashes).

also Sect. 4.2.1). Therefore, it assumes that the robot passed the T-junction and estimates the robot's position to be in the junction that forms a straight prolongation to the previous one. It is not until the robot has traveled some more distance before the generalization algorithm detects the corner (see Fig. 4.14d). Then, the position estimate is immediately corrected and a precise hypothesis is set up.

Such generalization delays cause most of the deviations between RouteLoc's position estimate and the robot's real world position. Please refer to the results chapter (Ch. 5) for details.

Precision of the Position Estimate. Because of the modeling of the environment and the robot's locomotion, RouteLoc is rather insensitive to odometry errors. Figure 4.15 shows the odometry data recorded during the "Big Route" experiment (for details, see Ch. 5). Please note the bad performance in measuring the

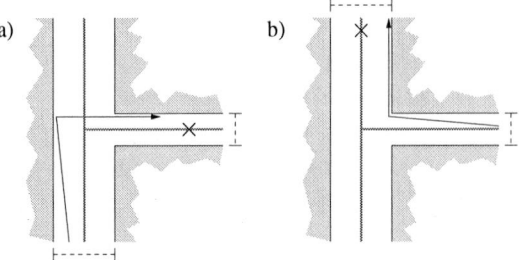

Figure 4.16: *Precision of the Position Estimate.*

angular component. The offsets normally represent only short distances that result from accumulating straight movements, and almost no rotational motion which often causes dead reckoning errors. Nevertheless, the precision of the algorithm is limited.

The precision is limited to half the width of the current corridor at right angles to the robot's driving direction and half the width of the previous corridor in the robot's driving direction (see Fig. 4.16). The figure shows two traverses of a T-junction: turning right from a wide corridor into a narrow corridor (Fig. 4.16a) and turning right from a narrow corridor into a wide corridor (Fig. 4.16b). The route graph is depicted in the center of each corridor as a grey line. The route traveled by the robot is shown as a thin black line. In both situations, the estimated position of the robot (marked by a cross on the route graph) differs from the real position of the robot (depicted by an arrow). Turning from a wide into a narrow corridor leads to a small horizontal error and a larger vertical error (Fig. 4.16a), whereas it is the other way round when turning from a narrow into a wide corridor (Fig. 4.16b).

The error could be even worse if the route graph is not correctly embedded in the center of the corridors, as it should be. Note that errors do not accumulate across junctions. However, within longer junctions odometry errors may become significant, because only corners provide a means to re-calibrate. But the experiments presented later give evidence that the error is acceptable.

The precision does *not* explicitly depend on the length of the traveled route, as every matching of a route corner to a graph junction once again limits the error. Nevertheless, the quality of the position estimate depends on the "quality" of the environment. The results of the different experiments presented in Ch. 5 support this point of view. For instance, the precision in narrow corridors is generally better because the robot almost *has to* travel on the route graph. In contrast to that, in open space it might deviate significantly.

4.4 Inside ROUTELOC: a Deeper Insight

The previous section has provided a complete overview of ROUTELOC. Nevertheless, some aspects are presented in a simplified way for clarity purposes. This section addresses the remaining problems and shows solutions in detail.

Section 4.3 used the unrealistic assumption that the route generalization algorithm creates a new corner for every decision point (junction) the robot passes, and—vice versa—that every generated corner has its counterpart in the route graph and thus in the real world. This is too optimistic, as will be shown below. Section 4.4.1 will deal with this problem and will present a general solution that requires no restrictive assumptions.

Right at the beginning of a robot journey, a few special cases have to be paid attention to: If the robot did not start its travel at a decision point but within a corridor, the standard matching process as described above does not work as fast as it could. Furthermore, a route with no real corner detected so far requires some special handling during the matching process. This will be discussed in Sect. 4.4.2.

Another assumption made in Sect. 4.3 is that the robot can only change its general driving direction at decision points. This is a straightforward inference from the definition of decision points (junctions) and corridors connecting these decision points. But there is a decision the robot can make anywhere, not only at decision points: turning around. Since the route graph junctions are directed, such a turning maneuver implies that the robot leaves the current junction. But unfortunately, it does not end in another junction represented in the route graph, because such turning junctions are not contained in the route graph as specified in Def. 4.2 on page 51. Section 4.4.3 describes the handling of turning around within corridors.

4.4.1 On Phantom Corners and Missed Junctions

While the robot travels, ROUTELOC is expected to ongoingly present a hypothesis about the robot's current position. This hypothesis is to be updated in regular intervals. In the experiments that will be presented in Ch. 5, an update interval of 20 cm travel distance is used. The number is arbitrarily chosen, it is no problem to substantially reduce the interval for example to every 5 cm, because the algorithm is very fast. But, as experiments have shown, a shorter update interval does not necessarily mean better results. In every update step, the route generalization algorithm checks whether a new corridor has been entered and updates the route description accordingly. Afterwards, the matching process is carried out which leads to a position estimate, as discussed in Sect. 4.3.4. A detailed description of ROUTELOC's control flow is presented in Sect. 4.4.5.

In every update step, four different situations can occur with respect to detected

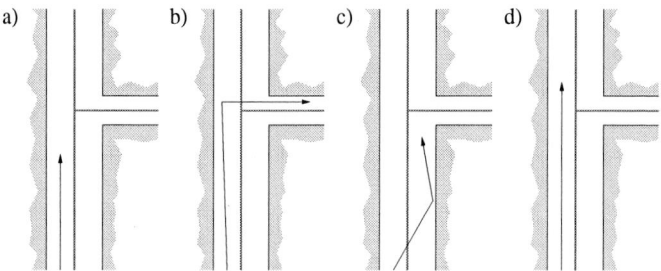

Figure 4.17: *Special Cases.* a) no junction, no corner; b) junction, corner; c) no junction, phantom corner; d) missed junction, no corner.

or not detected route corners on the one hand, and to existing or not existing junctions in the route graph on the other hand. The four sub-figures in Fig. 4.17 show a typical scenario: A long wide corridor that meets another, narrower corridor at a T-junction. The corresponding route graph fragment is depicted as grey lines in the middle of the corridors. The thin black arrows represent the generalizations of four different motion tracks that have been recorded by the robot's odometry.

1. There is no junction in reality and the generalization algorithm correctly detects no route corner (see Fig. 4.17a). This is the normal case because most of the time the robot moves through corridors without detecting any corners.

2. There is a junction in reality and the generalization algorithm correctly detects a corresponding route corner (see Fig. 4.17b). This is the assumption made in the previous section for every corner detection.

3. There is no junction in reality even though the generalization algorithm detects a route corner, a so-called *phantom corner* (see Fig. 4.17c). Unfortunately, this case is not that rare, because due to odometry drift, long corridors are often generalized to more than one segment.

4. There is a junction in reality but the route generalization algorithm does not detect a corresponding route corner (see Fig. 4.17d). This is the problem of *missed junctions* which is not a flaw of the route generalization algorithm but a result of the spartan sensor use of the approach. Nevertheless, RouteLoc is able to handle it.

   The correct handling of these four cases is fundamental for the algorithm. Therefore, they are discussed in the following subsections in detail.

Situation 1. **There is *no* Junction and *no* Corner is Detected.**

In Fig. 4.17a, the standard situation is illustrated: the robot moves within a corridor, no junction in its surroundings, and the route generalization algorithm correctly infers that the robot did not change corridors, but still travels in the same corridor as one step before. In this case, the matching process can be carried out as described in Sect. 4.3. There is only one restriction: the definition of the similarity measure in (4.3) on page 55 assumes that the final length of the route segment to be matched with the junction's outgoing segment is already known. As mentioned above, this is not the case for the currently final segment of the route traveled so far. Therefore, the calculation of the similarity measure $s_l(c, j)$ for the lengths of the final route corner c and a junction j has to be changed to

$$s_l(c, j) = \begin{cases} 1, & l_c \leq d_j \wedge c = c_n \\ sim\left(\frac{l_c - d_j}{d_j}, 0\right), & \text{otherwise} \end{cases} \qquad (4.12)$$

In (4.12), l_c is the length of the route segment of corner c; d_j is the length of the outgoing corridor of junction j. In contrast to the original definition in (4.3) on page 55, the similarity is set to 100% not only if the lengths are equal, but also if the final route segment is shorter than the junction segment. This is no surprise, as it is a preliminary match and the information about the final route segment available at that time indicates that it matches the route graph junction. Only if l_c eventually happens to be larger than d_j, the similarity measure drops below 100%. Note that (4.12) replaces (4.3) as definition of the similarity measure with respect to the segments' lengths. Such replacements of the definitions will often occur throughout the rest of this chapter.

As long as no corner is detected, there is no need for propagating the probabilities to adjacent junctions. Thus, the similarity values for each junction are only adapted to the current route generalization. Nevertheless, the case of missed junctions has to be kept in mind (see Situation 4 below).

Situation 2. **There is a Junction and a Corner is Detected.**

In some situations, the route generalization algorithm detects corners in the route, as depicted in Fig. 4.17b. If there exists a corresponding junction in the route graph, the matching as described in Sect. 4.3 will be successful. Note that detecting a new corner in the route fixes the then penultimate corner with respect to its angle and length components. Therefore, the matching is a three-step process in this case: first, the new penultimate corner is matched according to the rules described in Sect. 4.3.2 and the similarity measure $s_l(c_{n-1}, j)$ just defined in (4.12). Second, the probabilities are propagated to the adjacent junctions as discussed in

Sect. 4.3.3. And third, the new final corner c_n is matched as a non-fixed corner according to (4.12). This standard case was also discussed in Sect. 4.3.3.

Situation 3. There is *no* Junction, but a Corner is Detected.

As depicted in Fig. 4.17c, the generalized motion track as recorded by the robot's odometry can significantly deviate from a straight line even if the robot drives in a straight corridor. Especially in very long corridors, the odometry tends to be inaccurate. As an example, consider Fig. 5.5 on page 104 that depicts the generalized motion track that was recorded during experiments on the campus of the Universität Bremen. In the upper left part of the figure, the main boulevard of the campus, which is a straight, 190 m long part of the whole 300 m long boulevard, is partitioned into several segments. This is because the odometry recorded the straight boulevard as a crescent-shaped curve (see also the odometry data depicted in Fig. 4.15 on page 69). The erroneously detected "phantom corners" between the segments are a problem for RouteLoc because the matching qualities have to be propagated through the graph after every route corner detection (see the section on propagation, Sect. 4.3.3). If, however, such a detected route corner is a phantom corner, the propagation will be an error: Since no change of corridors has taken place, no propagation must be initiated.

Therefore, when detecting a corner, RouteLoc has to decide whether it is a corner with a corresponding junction in the route graph, or whether it is a phantom corner that results from bad odometry data. Note that such a decision is mainly based on uncertain knowledge, i. e. there is a remaining risk of a mistake. Please also note that for each junction the decision about whether or not the current corner is real may be different, i. e. the decision depends on the junction. As if this were not enough, this decision cannot be made until the information about the route corner is final. That means, the decision of whether or not a corner is believed to be either real or phantom can only be made with respect to the penultimate, already fixed, corner in the generalized route, and not with respect to the current corner. As a consequence, the classification of corner c_n has to be delayed until a new corner c_{n+1} is detected.

These considerations suggest to pursue two instead of one hypothesis for each junction (see Fig. 4.18). The left part of the figure shows the generalization of a route traveled by the robot as thin black lines (the arrow indicates the current robot position): a moment before a corner was detected (a), and a moment after the route corner was detected (b). The rightmost column shows the two corresponding hypotheses for a route graph junction that is depicted by grey lines. The question is whether the route corner detected in (b) is real or phantom. The first hypothesis describes the probability of the robot being in the outgoing segment of the junction and having been in the incoming segment before the final corner was detected (i. e.

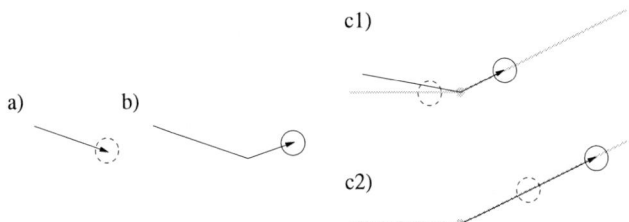

Figure 4.18: *Hypotheses for "Real" and "Phantom" Route Corners.*

the final route corner is *real*; see Fig. 4.18c1). The second hypothesis describes the probability of the robot being in the outgoing segment of the junction and already having been there before the final corner was detected (i. e. the final corner is *phantom*; see Fig. 4.18c2). The dashed circles mark the resulting position estimate for the hypotheses of the depicted junction and the given route before the corner was detected. The solid circles indicate the position estimate after the corner was detected. When a new corner is found, ROUTELOC chooses the more plausible of both hypotheses to be valid for the future. Note that this decision is irreversible, i. e. no history of hypotheses or the like is maintained.

As a result, two similarity measures for the two hypotheses have to be defined: The similarity measure that assumes the final route corner to be a *real* corner is identical to m_d as defined in (4.2) on page 55. Now, it is renamed to m_d^r for "direct matching quality assuming that the final route corner is real". The new similarity measure that assumes the final route corner to be a *phantom* corner is called m_d^p for "direct matching quality assuming that the final route corner is phantom". Analogous to (4.2), it is defined as

$$m_d^p(c, j) = s_l^p(c, j) \cdot s_\alpha^p(c, j). \tag{4.13}$$

If the corner c_n is assumed to be phantom, a new problem arises: in the matching process, the length l_c of the corresponding route segment is considered. But the value of l_c only represents the length of the route segment as determined by the generalization algorithm. However, if a corner is believed to be phantom, l_c also has to accumulate the previous corner's segment length. Furthermore, this might cascade over many corners. Therefore, ROUTELOC maintains a variable l_j^p for each junction that accumulates the length of those route segments the corresponding corners of which have been believed to be phantom. During the propagation process, l_j^p is set to the length of the penultimate corner's segment, if the corner is assumed to be real. l_j^p is increased by this length, if the corner is assumed to be phantom. Then, all calculations use $l_c + l_j^p$ instead of l_c if the "phantom" case is concerned. Thus, for each junction j, $l_j^p + l_c$ is the distance between the last corner believed in and the current robot position, assuming that the robot is currently located in j.

As a consequence, with respect to (4.3), the similarity measure s_l^p for the segments' lengths under the assumption that the last corner is phantom is defined as

$$s_l^p(c, j) = sim\left(\frac{|l_c + l_j^p - d_j|}{d_j}, 0\right) \tag{4.14}$$

The rotation angle similarity s_α^p for the "phantom case" is defined as

$$s_\alpha^p(c, j) = sim(\rho_c, 0) \tag{4.15}$$

In (4.15), the rotation angle ρ_c of the route corner is compared to $0°$, instead of being compared to the junction angle as in (4.4). As a result, the matching probability is close to 100% for very small angles (i. e. detected route corners with a small angle are likely to be phantom corners) and low for significant angles (i. e. detected route corners with an angle of, say, $90°$ are expected to be real corners with high probability).

The two hypotheses are always considered in parallel, i. e. there are two probabilities for a junction to host the robot: One of them assumes the final corner of route R to be a real corner, which means that the robot has been in the incoming segment of the junction before the corner was detected. The other one assumes that the final corner is a phantom corner, which means that the robot has already been in the outgoing segment of the junction before the corner was detected. As a result, there also exist two matching qualities $m^r(R, j)$ (assuming the final corner of R to be real) and $m^p(R, j)$ (assuming the final corner of R to be phantom) — based on the direct matching qualities $m_d^r(c, j)$ and $m_d^p(c, j)$ as defined above.

To determine the matching quality, these direct matching qualities *and* the following incoming matching qualities are required: The incoming matching quality $m_{in}(j)$ as introduced in (4.10) is renamed to $m_{in}^r(j)$ since it refers to the "real corner" case. As a complement, a new incoming matching quality $m_{in}^p(j)$ is defined that describes the probability that the robot has been in the incoming segment of j before the last corner had been detected under the assumption that this corner was phantom. That means, the robot has been in some junction j' with $j' \in in(j)$ before. During the propagation process, $m_{in}^p(j)$ is set according to the more plausible hypothesis:

$$m_{in}^p(j) = \begin{cases} m^r(R, j'), & m^r(R, j') \geq m^p(R, j') \\ m^p(R, j'), & \text{otherwise} \end{cases} \tag{4.16}$$

Finally, the overall probability of the junction (i. e. the matching quality) is then

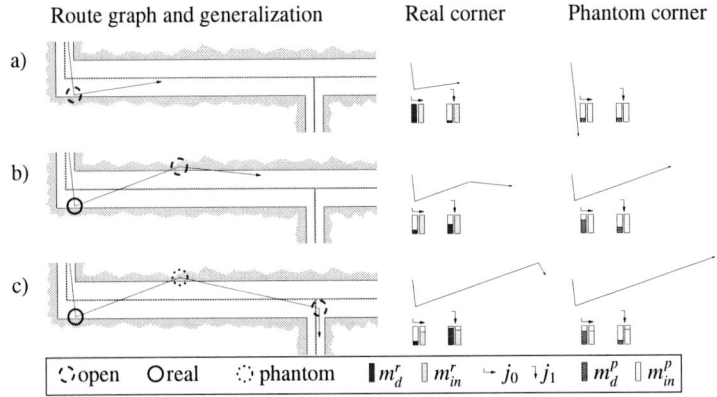

Figure 4.19: *Handling of Phantom Corners.*

calculated as the maximum of both hypotheses:

$$m^r(R, j) = m^r_{in}(j) \cdot m^r_d(c_n, j) \qquad (4.17)$$
$$m^p(R, j) = m^p_{in}(j) \cdot m^p_d(c_n, j)$$
$$m(R, j) = \max\{m^r(R, j), m^p(R, j)\}$$

To illustrate the handling of phantom corners, Fig. 4.19 shows a typical situation in a left turn: the robot enters a corridor and leaves it to the right after a while. The three rows depict three critical phases. From left to right, each row shows a part of the environment with the embedded route graph and the generalization of the route traveled so far (the arrow indicates the current robot position); the matching qualities for the assumption that the final corner is real; and the matching qualities for the assumption that the final corner is phantom. The "probability bars" refer to two junctions j_0 (left turn into the long corridor) and j_1 (right turn leaving the long corridor at the T-junction).

In Fig. 4.19a, the robot entered the long corridor after a left turn. The corresponding corner (highlighted by a dashed circle) is "open" for the time being, i. e. it cannot yet be decided whether it is real or phantom, because it is not yet fixed with respect to rotation angle and segment's length. The hypothesis "real corner" results in a perfect match for the junction j_0 and a poor match for j_1. Pursuing the assumption that the final corner is phantom means that the route should match a straight line of the length of the sum of both route segments. This yields only a poor direct match because the angle significantly deviates from 0° and (in case of j_1) the segment is longer than the junction's segment.

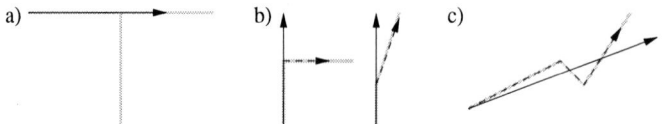

Figure 4.20: *Different Examples of Missed Junctions.*

In Fig. 4.19b, the route generalization algorithm detected a corner in the middle of the corridor. As a first result the former final corner can now be fixed with respect to length and rotation angle. As a consequence, a decision can be made, whether it should be regarded as real or phantom corner. For j_0, the hypothesis "real" is a clear winner (cf. Fig. 4.19a), thus the corner is declared to be real for j_0. Please note that the depicted circle labels attached to the route generalization always refer to junction j_0. As indicated by the probability bars, for j_1 it is slightly more likely that the corner is phantom than real. Similar to this decision, the hypothesis for the newly detected final corner shows that for j_0 the assumption "phantom" corner is more realistic. Instead, for j_1 a "real" corner seems more plausible. Overall, in Fig. 4.19b, j_0 is the candidate junction assuming the final corner to be phantom.

In Fig. 4.19c, another corner is detected directing the robot in a right turn to a new corridor. The decision about the then fixed penultimate corner can be made as already mentioned above (phantom for j_0 and real for j_1). As a result, the two hypotheses show a very good match for j_1 under the assumption that the final corner were real, thus j_1 is the candidate junction here.

Please note that the correct handling of phantom corners is a necessary feature but no panacea. It has to be kept in mind that each phantom corner causes a loss of information in that it may erroneously increase the matching quality of some short junctions: If a long straight corridor is generalized as a single segment, this segment would be too long to match any junctions. If, instead, the corridor is jigsawed by lots of short segments separated by phantom corners, they might match some short junction elsewhere in the route graph – at least directly.

Situation 4. There is a Junction, but *no* Corner is Detected

It is possible that the robot passes a corner existing in reality, while the route generalization algorithm does not (immediately) detect it. As a consequence, the resulting change of corridors is not recognized (*missed junction*). Usually, this cannot be blamed on the generalization but on the fact that—based only on proprioception—one cannot distinguish traveling in a straight corridor with no junctions or crossings from traveling in a straight corridor passing several T-junctions.

Figure 4.20 shows some examples of situations in which a junction has been passed by the robot without noticing. The route generalization is depicted as a thin black arrow whereas the route graph is shown as thicker grey lines. In Fig. 4.20a, a standard T-junction is depicted. The robot just passed the junction. However, the route generalization algorithm could not notice the change of junctions because the robot remains in the same real world corridor. Nevertheless, RouteLoc has to adapt the matching qualities adequately, i. e. the junction leading to the T has to drop in probability whereas its straight prolongation has to become more probable. In Fig. 4.20b, two situations are shown, where the route generalization algorithm misses a corner—at least for some time: Once again, the generalized route is depicted as thin black solid line. This time, the robot turns right as the dashed arrow indicates. The route generalization algorithm has not (yet) noticed the resulting change of corridors. This phenomenon has already been addressed under the label "generalization delay" in Sect. 4.3.4. Even though the involved junctions are not each others straight prolongations, RouteLoc has to infer that the robot changed junctions in such a case. In Fig. 4.20c, the route generalization algorithm failed to detect two route corners and thus the junctions were not recognized. It is RouteLoc's task to correctly map the route generalization (depicted as a solid arrow) onto the route graph and find the real position of the robot (dashed line), even though more than one junction has been missed. These three scenarios serve as a motivation for the discussion of the handling of missed junctions which is presented in detail throughout the rest of this section.

RouteLoc has to put some effort in the analysis of whether or not a junction has been passed (i. e. a change of corridors happened) even though no corner has been generalized. The first ingredient to this analysis is a means that helps to decide for a junction j if it actually has been overlooked by the route generalization algorithm. Given that the robot travels somewhere in the corridor represented by the outgoing segment of junction j with length d_j, a junction has been overlooked if the segment of the final route corner is longer than d_j. Two scenarios are plausible: First (this is the normal case), the final route corner may still match j but the matching quality drops since the length of the corner exceeds d_j. In this case, it is not junction j that has been overlooked but one of the adjacent junctions of which j is an incoming junction. Thus, this case can be handled here as usual. Second, the final route corner c_n may be an overlap from an *incoming* junction j' of j and c_n does not correspond to j but to j'. Both situations are depicted in Fig. 4.21. The route generalization algorithm provides RouteLoc with the final route corner as depicted in Fig. 4.21a. This route corner is to be matched with the left-turn junction that is highlighted by its labeling arrow in Fig. 4.21b. The route segment is longer than the junction's outgoing segment. Figure 4.21c shows the first interpretation: RouteLoc assumes that the corner matches j even though the segment is too long. As a result, the direct matching quality of j will drop, as can be seen

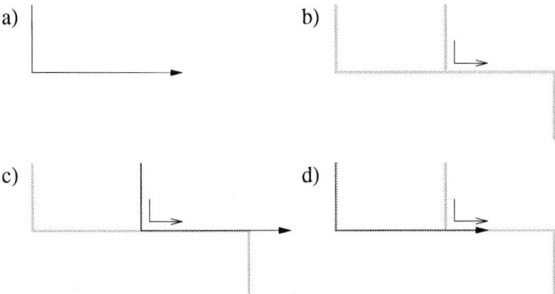

Figure 4.21: *The two Overlap Interpretations.*

in (4.3) and its subsequent refinements. The second interpretation is depicted in Fig. 4.21d, where RouteLoc infers that the final route corner corresponds to the junction j' with $j' \in$ in(j) and the route segment reaches into the outgoing segment of j. As a result, the current position estimate is somewhere in the outgoing segment of junction j.

Both scenarios have to be considered when updating the matching qualities of the junctions, but not in parallel as has been the case in other branches of RouteLoc. Here, a decision can immediately be made. RouteLoc only has to find out which of the scenarios is more likely. Determining the probability of the first interpretation (as depicted in Fig. 4.21c) is straightforwardly done according to the matching rules described in Sect. 4.3.2. However, for the second scenario, an "overlap matching quality" m_{ov} has to be defined: It consists of $m(R, j')$ with $j' \in$ in(j) *and* the probability p_{ov} of having overlooked the junction j. For junctions with a rotation angle of $0°$, so-called *straight junctions*, this probability is 1.0, but it drops depending on the angle deviation and on the length deviation. As usual, also here, two cases have to be taken into account: the assumption that the final corner is real (m_{ov}^r) and the hypothesis that the final corner is phantom (m_{ov}^p).

$$m_{ov}^r(R, j) = \max_{j' \in \text{in}(j)} \{m^r(R, j') \cdot p_{ov}\} \qquad (4.18)$$

$$m_{ov}^p(R, j) = \max_{j' \in \text{in}(j)} \{m^p(R, j') \cdot p_{ov}\} \qquad (4.19)$$

$$m_{ov}(R, j) = \max\{m_{ov}^r(R, j), m_{ov}^p(R, j)\} \qquad (4.20)$$

Please note that the calculation of the maximum is necessary to choose the best matching overlap from all incoming junctions. This is shown in Fig. 4.22. Figure 4.22a depicts the generalization of the current route segment. Figs. 4.22b-d visualize the possible assignments to a sample route graph. If it turns out that

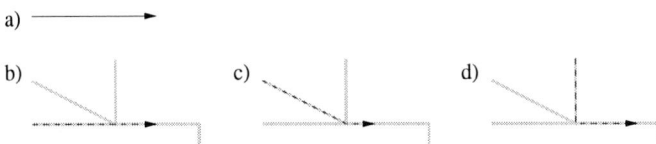

Figure 4.22: *Determining the Most Probable Overlap.*

$m_{ov}(R, j)$ (i.e. the matching quality for the scenario depicted in Fig. 4.21c) is greater than $m(R, j)$ (matching quality for the scenario depicted in Fig. 4.21d), RouteLoc favors the scenario that the route generalization algorithm has failed to detect a corner and thus RouteLoc missed junction j. Due to the best match in the angular component, the case shown in Fig. 4.21 would be preferred here.

The probability $p_{ov}(R, j)$ of having overlooked junction j is defined as

$$p_{ov}(R, j) = f_{l_c^+} \cdot p_{ov}^\alpha(R, j) \cdot p_{ov}^\theta(R, j) \cdot p_{ov}^b(R, j) \qquad (4.21)$$

As shown in (4.21), the value of $p_{ov}(R, j)$ depends on four different factors:

Factor $f_{l_c^+}$: The longer the overlap, the more probable it is having overlooked a junction. $f_{l_c^+}$ is a reinforcement factor that stretches or compresses the probability distribution. It depends on the length l_c^+ of the overlap into junction j (see below).

Factor $p_{ov}^\alpha(R, j)$: The closer the junction angle to zero, the more likely it is having overlooked the junction. The angular component of $p_{ov}(R, j)$ is defined as

$$p_{ov}^\alpha(R, j) = sim(k_1 \gamma_j, 0) \qquad (4.22)$$

As mentioned, for straight junctions, $p_{ov}^\alpha(R, j) = 1.0$. Besides, k_1 is a constant weighting factor.

Factor $p_{ov}^\theta(R, j)$: The relative deviation between the robot's current orientation in the world and the orientation of the corridor it is currently traveling in is also taken into account. The closer this orientation deviation to the value of the rotation angle of the junction, the more probable it is that the junction has been overlooked. Then, the robot has probably already turned into the new corridor. The orientation component of $p_{ov}(R, j)$ is defined as

$$p_{ov}^\theta(R, j) = sim(k_2 \gamma_j, \theta) \qquad (4.23)$$

k_2 is a constant weighting factor.

Factor $p_{ov}^b(R, j)$**:** Junctions that have only few successive junctions result in a higher overlap probability. This is because it is more probable that there is an overlap to junction j from junction k if j is the only potential successor of k. The branching component of $p_{ov}(R, j)$ is defined as

$$p_{ov}^b(R, j) = sim(k_3 \cdot (b - 1), 1) \tag{4.24}$$

b is the number of possible transitions from junction j. k_3 is a constant weighting factor.

As shown in (4.22)-(4.24), determining the three probabilities $p_{ov}^\alpha(R, j)$, $p_{ov}^\theta(R, j)$, and $p_{ov}^b(R, j)$ is straightforward; calculating f_j^+ is slightly more difficult. If RouteLoc had to consider only overlaps across straight junctions, the calculation of the overlap length would be directly possible, namely the difference of the route segment's length and the incoming junction's segment length. But, as depicted in Fig. 4.20b, missed junctions "are allowed" to have arbitrary rotation angles. Therefore, RouteLoc has to provide a general solution to this problem.

Figure 4.23 shows the known lengths and angles on the one hand, and the values to be determined on the other hand. A part of a route graph is depicted as thick grey lines. Two of its junctions are relevant here: j' and j. They are marked by their labeling arrows. The generalized route is shown as a dashed line and embedded in the route graph. It overshoots the outgoing segment of junction j'. Therefore, RouteLoc assumes that the route generalization algorithm failed to detect a corner which indicated the change of corridors into junction j. It is RouteLoc's task to determine the *overlap length* o_j^+. To calculate o_j^+, the law of sines is used. It states that in each triangle the ratio of a triangle's side and the sine of its opposite angle is equal for every side:

$$\frac{\sin \alpha}{a} = \frac{\sin \beta}{b} = \frac{\sin \gamma}{c} \tag{4.25}$$

By further exploiting the fact that a triangle's angles sum up to π, the unknown overlap length can be determined as:

$$l_c^+ = \begin{cases} \dfrac{l_c \cdot \sin\left(\pi - \gamma_j - \arcsin\left(\frac{d_{j'} \cdot \sin \gamma_j}{l_c}\right)\right)}{\sin \gamma_j} & \text{, assumption } c_n \text{ is real} \\[4ex] \dfrac{(l_c + l_j^p) \cdot \sin\left(\pi - \gamma_j - \arcsin\left(\frac{d_{j'} \cdot \sin \gamma_j}{l_c + l_j^p}\right)\right)}{\sin \gamma_j} & \text{, assumption } c_n \text{ is phantom} \end{cases} \tag{4.26}$$

Up to here, the scenarios shown in Fig. 4.20a and Fig. 4.20b can be solved. But as already indicated in Fig. 4.20c, the overlap may stretch over more than one junction. For instance, in a long straight corridor, it is likely that many rooms may be entered from this corridor. In such a situation, the route graph would look like the one depicted in Fig. 4.24b. It can be expected that the generalization algorithm will generate a route corner similar to the one depicted in Fig. 4.24a. As

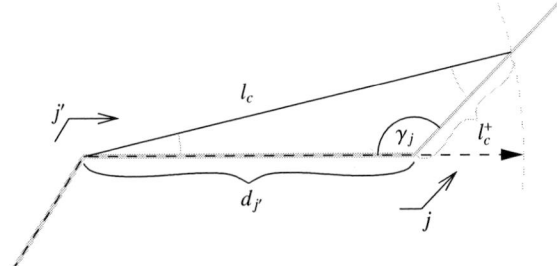

Figure 4.23: *Calculating how far the Final Route Segment Extends Into the Currently Considered Junction's Outgoing Segment.* The items l_c, $d_{j'}$ and γ_j are given, l_c^+ is to be determined.

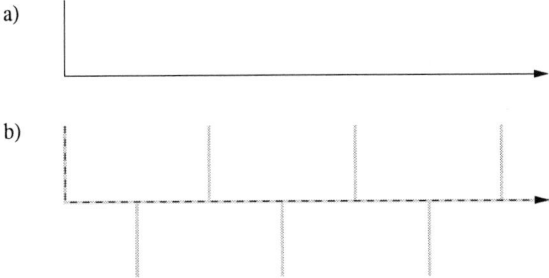

Figure 4.24: *The Overlap Stretches Over Many Junctions.* a) generalization of final route corner; b) part of a route graph with embedded generalized route.

indicated by the dashed arrow in the route graph (see Fig. 4.24b), it is necessary to take overlaps into account that stretch across many intermediate junctions into the candidate junction. Thus, a change in (4.26) is required: The distance $d_{j'}$, i.e. the distance from the incoming junction j' to the currently considered junction j is only helpful, if the overlap originates from an incoming junction of j. But in the general case, the overlap may stretch over a sequence of junctions. Therefore, the relevant distance is the Euclidian distance between the route graph position Pos_{c_n} of the last detected corner c_n and the position Pos_j of the current junction j. Therefore, the equations have to be specified more generally by substituting $d_{j'}$ with the Euclidian distance $|Pos_{c_n} - Pos_j|$ between the route graph position of the last generalized corner and the currently considered junction. Fortunately, the route graph position of the last generalized corner can easily be determined during the propagation process. As a consequence, every calculation has to consider l_c^+

instead of l_c and $l_c + l_j^p$, respectively. And it has to use $d_j^+ := d_j + |Pos_{c_n} - Pos_j|$ instead of d_j. See also Sect. 4.4.4.

At this point, RouteLoc is able to decide whether there is an overlap from a previous junction that stretches into the currently considered junction j. But what has to be done, if such an overlap situation is detected? If there is an overlap into junction j from one of its incoming junctions, say j', j temporarily "inherits" the characteristics of j'. That means, that j is treated as if it were j' until a new corner is generalized. Please note that this inheritance process may cascade across a sequence of adjacent junctions that are "bridged" by the overlap that stretches into j.

As a result, the matching quality of j is correctly set in an overlap situation, even though the route segment length does not fit into the outgoing segment of j as such. But by taking into account the possibility that this route segment may have already been started in one of j's incoming junctions, the situation can be clarified. The temporarily changed junction properties are re-set to their original values if a new route corner is detected.

To summarize, the route generalization algorithm is sometimes not able to detect junctions due to the minimal sensor use. RouteLoc is able to solve the problem of "missed junctions" by checking whether overlaps from previous junctions could host the current route segment.

4.4.2   Initial Phase Details

After having solved the phantom corner and missed junction problems in section 4.4.1, there are two situations that deserve special handling during the early phases of a robot journey. These have to be covered by RouteLoc, but have not been addressed yet:

- Matching a route $R = \langle c_o \rangle$ that comprises only the initial corner with the route graph, and

- starting the robot's journey not at a decision point but somewhere in the middle of a corridor.

These two topics are discussed in the following two paragraphs.

Before the First Corner was Detected.   As discussed in Sect. 4.2.1, the rotation angle of the initial route corner c_0 is special in that it is a "don't care" value. Further more, it must never be used during the matching process, because it has no meaning: it describes the rotation angle between the first route segment and an imaginary "zeroth" route segment. Therefore, the matching process has to be carried out slightly differently as long as no real route corner has been detected.

The implementation of this requirement is straightforwardly achieved by a further extension to the similarity measure calculation previously shown in (4.4) and refined in (4.15). The equation that includes the "before the first corner" case looks as follows for the assumption that c_n is a real corner:

$$s_\alpha^r(c, j) = \begin{cases} 1, & c = c_0 \\ sim\left(\|\gamma_j - \rho_c\|, 0\right), & \text{otherwise} \end{cases} \qquad (4.27)$$

and for the assumption that c_n is phantom:

$$s_\alpha^p(c, j) = \begin{cases} 1, & c = c_0 \\ sim\left(\rho_c, 0\right), & \text{otherwise} \end{cases} \qquad (4.28)$$

where c_0 is the initial corner of the route.

If the route R only comprises one corner (the "don't care corner"), i.e. $R = \langle c_0 \rangle$, the angle is ignored, because it is the initial rotation angle that has no meaning (cf. page 46), thus $s_\alpha(c_0, j) = 1$. Therefore, the only remaining criterion for a direct match is the segment's length, thus $m_d^r(c, j) = s_l^r(c, j)$ and $m_d^p(c, j) = s_l^p(c, j)$ in this case.

Starting in the Middle of a Corridor. The basic idea of the whole approach is that detected route corners can be identified with certain junctions in the route graph. Then, the similarity measures deliver an adequate means to decide about the matching quality. However, at the very beginning of a robot journey, it will often happen that the robot does not start at a place in the real world that is represented by a route graph node. Instead, the starting position could be located somewhere in a corridor in the middle between two decision points. If the robot reached the first adjacent junction, detected a corner and matched the route with the graph, the length of the driven segment would be significantly too short in comparison with the junction's outgoing segment (because the robot started in the middle). Nevertheless, the route segment perfectly fits into the route graph. Thus, the first route segment must be allowed to be shorter than the junction's outgoing segment without loss of matching quality. Once again, the equation for the similarity measure is refined to:

$$s_l^r(c, j) = \begin{cases} 1, & l_c^+ \le d_j \wedge c \in \{c_0, c_n\} \\ sim\left(\frac{l_c^+ - d_j}{d_j}, 0\right), & \text{otherwise} \end{cases} \qquad (4.29)$$

$$s_l^p(c, j) = \begin{cases} 1, & l_c^+ \le d_j \wedge c \in \{c_1, c_n\} \\ sim\left(\frac{l_c^+ - d_j}{d_j}, 0\right), & \text{otherwise} \end{cases} \qquad (4.30)$$

where c_0 is the "don't care corner" and c_1 is the first real corner of the route $R = \langle c_0, c_1, \ldots, c_n \rangle$.

For s_l^p, c_1 is also allowed, because the previous corner was not believed in.

4.4.3   Turning Around Within a Corridor

Nonholonomic vehicles such as the Bremen Autonomous Wheelchair "Rolland" are not able to move in arbitrary directions. Instead, they are restricted to bias bearings such as forwards and backwards. As a consequence, nonholonomic robots are not able to turn on the spot without shunting. Indeed for the wheelchair, there are some corridors that are too narrow to turn in at all. Therefore, it is fundamental to know the orientation of the wheelchair within a corridor. This is solved by modeling the corridors as one-way junctions, where the orientation is inherently known (see Sect. 4.2.2 on route graphs). If the robot turns around in a corridor, it leaves its current junction even though it remains in the same corridor.

But—by definition—leaving a junction means to enter another junction. Unfortunately, there are no junctions in the route graph that connect the two directions of a corridor.

An additional problem is that a turning maneuver can be carried out at any position within the corridor. In contrast to that, leaving the corridor is only possible at junctions. As a consequence, maintaining an offset of how far the robot already traveled in a junction is not suitable since the "origin" of a turn-junction is variable, depending on the point in the corridor where the robot actually performed the turning maneuver.

To overcome these problems in order to be able to handle turns, the set of junctions that initially form the route graph G is extended by so-called *turn-junctions* at program start as follows:

$$
G' = G \cup \left\{ \left(H, T, \pi, |\overline{HT}|, I \right) \middle| \right.
$$
$$
\left. H, T \in V, I \subseteq G, \forall i \in I : i = (T, H, \gamma_i, |\overline{TH}|, I') \right\} \tag{4.31}
$$

In (4.31), for each junction j_i in the initial route graph G, all turn-junctions that can be generated for j_i are added to G. H and T are elements of the set V of nodes of the route graph. As an example, consider the route graph depicted in Fig. 5.4b that is used for the experiments presented in Ch. 5. The 144 junctions of this route graph require an additional set of 102 turn-junctions. The upper bound of the number of required turn-junctions for a route graph with n "real" junctions is $2n$. In typical environments, however, it often happens that two or more junctions share one turn-junction, e. g. junctions **cdh** and **kdh** in Fig. 5.4b both need the turn-junction **dhd**. The incoming and the outgoing segment of these turn-junctions represent the same corridor (forwards and backwards direction) and have a rotation angle of $180°$. After having generated the turn-junctions at program start, they are dealt with as if they were "normal" junctions from that point.

The special handling of these turn-junctions has to be considered at various points within ROUTELOC:

The "turn offset". As mentioned above, a turn-junction must store the information concerning where the robot actually turned within the corresponding corridor. Therefore, each junction in the route graph is assigned with a variable which represents the estimated distance between the turning position and the origin of the junction. For non-turn-junctions, the turn offset is set to zero.

Calculation of Matching Qualities. When calculating the matching quality of a generalized route corner with such a turn-junction, an undershooting is granted: the deviation of the length is accepted if the route segment is too short. If the route segment is too long, the matching quality will drop.

Missed Junctions. It is assumed that turns are not overlooked by the generalization algorithm. This is why turn-junctions do not have to be considered when the check for missed junctions is performed as described in Sect. 4.4.1. But if a junction temporarily has to adopt the data from an incoming turn-junction, its position cannot be set to the turn-junctions' original position but to the estimated position of the turn.

Auxiliary Functions. In some auxiliary functions, the special case of a turn-junction has to be considered, e. g., when calculating the distance that has already been traveled in a junction (overlap from previous junctions).

As a consequence of the first item mentioned, the similarity measure definitions for the lengths has to be revised once more:

$$s_l^r(c, j) = \begin{cases} 1, & l_c^+ \leq d_j \wedge (c \in \{c_0, c_n\} \vee \text{isTurn}(j)) \\ sim\left(\frac{l_c^+ - d_j}{d_j}, 0\right), & \text{otherwise} \end{cases} \qquad (4.32)$$

$$s_l^p(c, j) = \begin{cases} 1, & l_c^+ \leq d_j \wedge (c \in \{c_1, c_n\} \vee \text{isTurn}(j)) \\ sim\left(\frac{l_c^+ - d_j}{d_j}, 0\right), & \text{otherwise} \end{cases} \qquad (4.33)$$

where isTurn(j) equals *true*, if and only if j is one of the artificial turn-junctions.

Figure 4.25 illustrates the handling of turn-junctions. The rows a) – c) show three snapshots of a turning maneuver. The generalized route is depicted on the left as a thin black line, the arrow indicates the current position of the robot. The right part of each row shows a part of the route graph that is annotated with the junctions probability values. As already known from other figures, the left bar (filled with dark grey) represents the direct matching quality, whereas the right bar (filled with light grey) stands for the incoming matching quality. The small arrow next to the probability bars indicates the junction they belong to. Note that

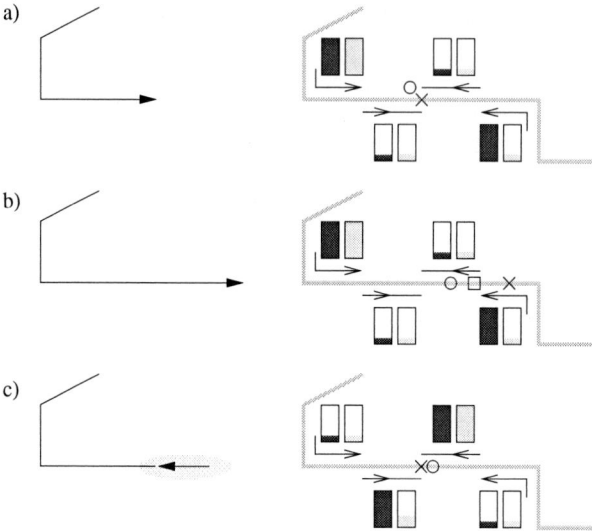

Figure 4.25: *Turn-Junctions.*

these figures only show the hypothesis that the current corner is a real corner. In Fig. 4.25a, the robot just conducted a 90° left turn. The matching qualities indicate that ROUTELOC correctly associates this corner with the junction in the upper left corner of the route graph. Even though the direct matching quality of the junction in the lower right part of the route graph also delivers a perfect match, its incoming matching quality is rather poor. The turn-junctions for either direction of the main corridor are not very likely. The real robot position is indicated by the circle, the position estimated by ROUTELOC is marked by a cross. The position estimate is almost correct.

After some more distance in the main corridor, the robot turns around, i. e. it performs a 180° turn on the spot within the corridor (turning position indicated by a square in the Fig. 4.25b). This is not immediately recognized by the generalization algorithm, as shown in Fig. 4.25b (see also the section on the generalization delay, Sect. 4.3.4). As a consequence, the hypotheses do not change and ROUTELOC still assumes that the robot follows the main corridor. The estimated position (marked by the cross) deviates significantly from the real position (marked by the circle).

As soon as the route generalization algorithm detects the turning maneuver

(Fig. 4.25c), the situation changes: The robot is still in the main corridor, but it now drives in the opposite direction. If there were no turn-junctions, the 90°-left junction in the lower right part of the route graph would have to host the robot then. But this would be an error since neither the direct nor the incoming matching quality give good matches. Nevertheless, due to the introduction of the turn-junctions, the situation can easily be clarified. The position estimate is once again very close to the real position.

4.4.4 Similarity Measures (Revised)

This section recapitulates the defining equations for the similarity measures including all special cases. First, the equations are shown for the assumption that the final corner is a real corner:

$$m_d^r(c, j) = s_l^r(c, j) \cdot s_\alpha^r(c, j) \tag{4.34}$$

$$s_l^r(c, j) = \begin{cases} 1, & l_c^+ \leq d_j \wedge (c \in \{c_0, c_n\} \vee \text{isTurn}(j)) \\ sim\left(\frac{l_c^+ - d_j}{d_j}, 0\right), & \text{otherwise} \end{cases} \tag{4.35}$$

$$s_\alpha^r(c, j) = \begin{cases} 1, & c = c_0 \\ sim\left(\|\gamma_j - \rho_c\|, 0\right), & \text{otherwise} \end{cases} \tag{4.36}$$

For the hypothesis that the final corner is not existing in reality ("phantom corner"), the equations are defined as follows:

$$m_d^p(c, j) = s_l^p(c, j) \cdot s_\alpha^p(c, j) \tag{4.37}$$

$$s_l^p(c, j) = \begin{cases} 1, & l_c^+ \leq d_j \wedge (c \in \{c_1, c_n\} \vee \text{isTurn}(j)) \\ sim\left(\frac{l_c^+ - d_j}{d_j}, 0\right), & \text{otherwise} \end{cases} \tag{4.38}$$

$$s_\alpha^p(c, j) = \begin{cases} 1, & c = c_0 \\ sim\left(\rho_c, 0\right), & \text{otherwise} \end{cases} \tag{4.39}$$

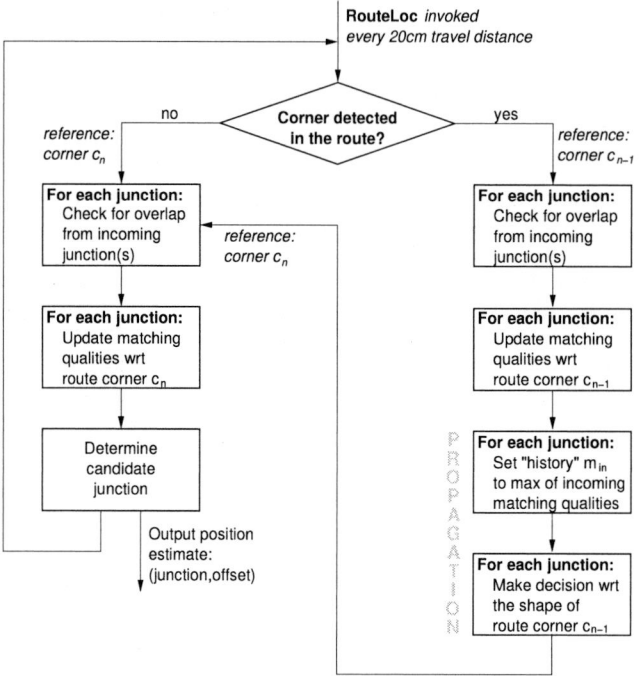

Figure 4.26: *Control Loop of* RouteLoc.

4.4.5 Control Flow

This section explains the control flow of the RouteLoc implementation used for the experiments presented in Sect. 5 below. Figure 4.26 depicts the algorithm's main control loop.

Apart from the a priori knowledge (the route graph of the environment), RouteLoc expects the segment length and the rotation angle of the penultimate (and already fixed) as well as those of the (not yet fixed) final corner of the generalized route as input. Furthermore, it requires the information of whether a change of corridors has taken place. As described in Sect. 4.3.4, RouteLoc uses this information to calculate a position estimate which is given in the form "The robot's position is in corridor **XY**, 123 cm after passing **X**.".

Since RouteLoc is embedded in the software architecture of the wheelchair robot "Rolland" as discussed in Röfer and Lankenau (2000), ROUTELOC receives new input data every 32 ms. Due to its low complexity, it is able to output a position estimate in the same cycle. Nevertheless, it would not be helpful to stick to this rapid cycle since the wheelchair travels only 2.7 cm at maximum speed in each cycle. Therefore, the current implementation uses a discretization of 20 cm; i. e. every 20 cm travel distance, ROUTELOC is invoked to determine the robot's location.

If the route generalization module decides that the robot has not left the current corridor (i. e. no corner detected and thus no new route segment started), ROUTELOC updates the matching quality of each junction in relation to the current corner c_n of the route (cf. the left branch depicted in Fig. 4.26). Even though the robot remains in the same corridor, an update is indispensable since the length and the rotation angle of the current corner may have changed. As a result, the matching qualities also change, overlaps may have to be considered, etc.

If instead, the route generalization module detects a new route corner, ROUTELOC makes the final decision about the previous route corner c_{n-1} with respect to angle and segment length (cf. the right branch depicted in Fig. 4.26). The matching probabilities are propagated to adjacent junctions. And finally, an update of the matching qualities is carried out with the new final corner c_n as reference corner. This process is presented in detail in Sect. 4.3.3.

No New Route Corner Detected

If the route generalization algorithm does not detect a new route corner, ROUTELOC conducts a normal update step as shown in the following pseudo-code fragment. Please note that the abstraction level of the code is chosen such that every function call is of constant complexity.

```
─────────────── No Corner Detected ───────────────
 1  refC := c_n
 2  FOR j IN Junctions DO
 3    max := m(R,j)
 4    FOR j' IN in(j) DO
 5      IF (checkOverlap(j',j) AND p_ov(refC,j)>max) THEN
 6        max := p_ov
 7        copyProperties(j',j)
 8      FI
 9    OD
10  OD

11  FOR j IN Junctions DO
12    calcMatchingQuality(j)
13  OD
```

First, ROUTELOC has to decide for each junction whether it hosts an overlap from one of its incoming junctions (see lines 2–10). Therefore, the "missed junction" procedure as described in *Situation 4* in Sect. 4.4.1 has to be carried out for each junction. As a result, some junctions may temporarily adopt the properties of one of their incoming junctions. Note that the set Junctions referred to in the code includes the "normal" junctions as well as the turn-junctions which are constructed at program start (see Sect. 4.4.3). Finally, the matching quality of each junction is determined based on the current junction properties. Please note that all the calculations referred to here are done for the "real corner" as well as for the "phantom corner" hypotheses, even though this is not made explicit neither in the above pseudo-code nor in the following code fragments.

New Route Corner Detected

If the route generalization algorithm announces that a new corner c_n has been detected, ROUTELOC has to do three things: First, the new penultimate corner c_{n-1} is fixed with respect to its rotation angle and length. Therefore, ROUTELOC performs the usual update step with this corner as reference corner (see lines 2–10 in the following pseudo code fragment).

```
─────────────────── Corner Detected ───────────────────
 1  refC := c_n-1
 2  FOR j IN Junctions DO
 3    max := m(R,j)
 4    FOR j' IN in(j) DO
 5      IF (checkOverlap(j',j) AND p_ov(refC,j')>max) THEN
 6        max := p_ov
 7        copyProperties(j',j)
 8      FI
 9    OD
10  OD

11  FOR j IN Junctions DO
12    calcMatchingQuality(j)
13  OD
```

Then, the matching qualities for all junctions are up-to-date, as if the route ended with c_{n-1}. Due to the fact that a new final corner c_n has been detected, the matching qualities have to be propagated to the adjacent junctions, as described in Sect. 4.3.3.

This is done by the loop in lines 14–21 in the code fragment shown here:

```
——————————— Corner Detected (cont'd.) ———————————
14  FOR j IN Junctions DO
15    max := m(R,j)
16    FOR j' IN in(j) DO
17      IF (m_in(R,j)>max) THEN
18        max := m_in(R,j)
19      FI
20    OD
21  OD

22  FOR j IN Junctions DO
23    decidePhantomVsReal(j)
24    resetTemporaryChanges(j)
25    normalizeMatchingQuality(j)
26  OD
```

According to Sect. 4.3.3, the propagation of the incoming matching qualities is followed by the decision about whether route corner c_{n-1} is finally assumed to be a real or a phantom corner. Besides, the temporary changes made during the handling of missed junctions have to be reset. As a result, the properties (rotation angle, last corner position, etc.) of each junction are restored to their original values. Furthermore, the matching qualities are normalized (lines 22–26). Finally, each junction has to be matched with the new final route corner c_n with the usual update strategy as shown in the following code fragment.

```
——————————— Corner Detected (cont'd.) ———————————
27  refC := c_n
28  FOR j IN Junctions DO
29    max := m(R,j)
30    FOR j' IN in(j) DO
31      IF (checkOverlap(j',j) AND p_ov(refC,j)>max) THEN
32        max := p_ov
33        copyProperties(j',j)
34      FI
35    OD
36  OD

37  max := 0
38  FOR j IN Junctions DO
39    CalcMatchingProbs(j)
40    IF (m(R,j) > max) THEN
41      max: = m(R,j)
42      candidateJunction := j
43    FI
44  OD
```

Table 4.2: ROUTELOC's *Computational Complexity.*

No.	Procedure	Complexity
1	`Overlap Check`	$O(n \cdot m)$
2	`Update Matching Qualities`	$O(n)$
3	`Propagation`	$O(n \cdot m)$
4	`"Housekeeping"`	$O(n)$
5	`Overlap Check`	$O(n \cdot m)$
6	`Update Matching Qualities`	$O(n)$
7	ROUTELOC	$O(n \cdot m)$

As a result, ROUTELOC determines the candidate junction by searching the set of junctions for the best matching junction (lines 37–44).

The above code fragments are a more detailed version of Fig. 4.26. They are presented here in order to clarify the discussion about the computational complexity of ROUTELOC in the following subsection.

4.4.6   Some Thoughts about Complexity

Within this subsection, the computational complexity of ROUTELOC is determined by analyzing the above code fragments.

ROUTELOC is embedded in the software architecture of the Bremen Autonomous Wheelchair "Rolland". Even though the wheelchair's modular setup is able to incorporate non-real time modules, such as complex image processing tasks or the like, the self-localization process should be fast and capable for real-time processing, i. e. it should be guaranteed that its answer is delivered within a bounded period of time.

A first milestone to reach this goal is the observation that the complexity of ROUTELOC does not at all depend on the length of the route, neither metric nor with respect to the number of detected corners. This is because of the inductive matching approach that guarantees that only two corners of the route are considered in ROUTELOC at a time, namely the penultimate and the current corner. As a result, matching a route comprising a thousand corners takes exactly as long as matching a single segment with the route graph.

The decisive factor for determining the complexity of ROUTELOC is the number of junctions represented in the route graph. As shown in Sect. 4.4.5, almost all calculations have to be performed for each junction in every update step. Table 4.2 lists the complexity of the specific parts of ROUTELOC as depicted in Fig. 4.26.

In Tab. 4.2, n denotes the number of junctions in the route graph (including the artificial turn-junctions that have once to be generated at program start). The variable m refers to the maximum number of incoming junctions of a junction. The overlap check and the propagation need to process two nested loops, the outer one runs through the set of junctions while the inner one covers all incoming junctions for each junction. Please note that the current implementation of the overlap check is probably not optimal. It simply reiterates the overlap check procedure a constant number of times in order to ensure that cascading overlaps from remote junctions are correctly considered. Since the algorithm is fast enough, this "brute-force" method is suitable here, but it should be replaced by a better solution in the future. By the way, it is reasonable to assume that a small number of iterations, say 10, is enough to correctly cover the cascading overlaps. This is because it is rather unlikely that the robot passes more than 10 junctions without detecting any corner, because also phantom corners would "help" here.

As a result, the computational complexity of these algorithm parts is $O(n \cdot m)$. Thus, ROUTELOC's overall complexity is found to be also $O(n \cdot m)$. But how many junctions (n) are there? And what is their maximum number (m) of incoming junctions? As discussed above, the number of junctions is determined by the number of decision points in the real world and corridors between them. The number of junctions does explicitly *not* depend on the spatial extent of the environment. Since the O-notation requires to consider the worst case, it is assumed that the route graph represents a completely connected network of corridors. This means that any two decision points are connected by a corridor. Please note that this is merely a gedankenexperiment, the practical view is discussed below.

Given that the completely connected corridor network comprises d decision points, there are $c = d(d - 1)/2$ corridors and $r = d(d - 1)(d - 2)$ "regular junctions". At program start, these regular junctions are complemented by $t = 2c$ turn-junctions. Thus, the total number of junctions amounts to $j = r + t = d(d - 1)(d - 2) + 2 \cdot d(d - 1)/2 = d(d - 1)^2$. Each of the j junctions has a total number of $i = d - 1$ incoming junctions, exactly one of which is a turn-junction. As a result, the theoretic computational complexity of ROUTELOC is $O(d(d - 1)^2 \cdot (d - 1)) = O(d^4)$.

But, as mentioned, this is only a theoretic value without any practical relevance. Even from a theoretic point of view the number can be drastically reduced by making the reasonable assumption that corridors may only intersect at decision points, i. e. the graph is planar. According to Euler's formula $E - V = F - 2$ about the relation of the number of vertices (V), edges (E) and faces (F) in planar graphs, the maximum number of corridors in a route graph significantly drops to $c = 3d - 6$. Hence, in the worst case scenario (a star-like scenario where any two corridors form a junction), the number of "regular junctions" in a planar route graph is equal to $j = c(c - 1)/2 = (3d - 6)(3d - 7)/2 = (9d^2 - 39d + 42)/2$. Since it is rather unlikely that the worst case route graph (with respect to the number of required junctions) is a valid representation of the environment, this remains a theoretic value.

In practice, it is very reasonable to expect the maximum number of possible incoming junctions a small constant number, say 100. Thus a hundred corridors have to meet at a certain decision point, which is already rather unlikely. By demanding such a restriction of incoming junctions, the *realistic* computational complexity of RouteLoc is reduced to $O(j)$.

Due to the fact that the number of junctions is usually related sub-linearly (or linearly at most) to the size of the environment, the approach scales very well. It would be interesting to test it in very large environments such as whole cities. As an example, consider Berlin. According to Behrens (2002), the digital street map of Berlin offered by the TeleAtlas AG comprises around 56000 line segments (corresponding to RouteLoc's corridors) and around 39000 decision points (nodes in the route graph). In contrast to the corridors represented in a route graph, the line segments in the digital street map may be curved. The line segments are annotated with attributes such as allowed speed, one-way information, etc. About every second line segment hosts an attribute change. As a side remark, the resolution of the digital map is 5 m in the inner city area and about 10 m in the outer area. This is interesting in the light of the performance evaluation of RouteLoc in Ch. 5.

5

A theory is something nobody believes, except the person who made it.
An experiment is something everybody believes, except the person who made it.

Albert Einstein

ROUTELOC: Experimental Results and Assessment

This chapter evaluates the performance of ROUTELOC in real world experiments. The Bremen Autonomous Wheelchair is used as a robotic platform to travel along the different routes. First, two testbeds are briefly presented: a middle-scale indoor environment and a large-scale hybrid indoor and outdoor environment. In Sect. 5.2, the evaluation method is discussed. Sect. 5.3 shows the results of the evaluation of the experiments. Finally, a conclusion is drawn and some perspectives for future work are discussed (see Sect. 5.4).

5.1 Experimental Setup

Experiments with the Bremen Autonomous Wheelchair "Rolland" were carried out on the campus of the Universität Bremen. Two different settings are discussed in the following subsections. The scenario for ROUTELOC's primary performance check is presented in Sect. 5.1.1, an indoor- and outdoor environment which covers a large part of the campus of the Universität Bremen. As a complement, Sect. 5.1.2 presents an indoor scenario in the third floor of the MZH building.

5.1.1 "The Big Route" – 2,176 m Across the Campus

During the "Big Route" experiments the wheelchair drove indoors and outdoors, visited seven different buildings and passed the boulevard which connects the buildings. Figure 5.1 shows a sketch of the campus of the Universität Bremen. The depicted area extends over about 380 m × 322 m. The buildings visited by the wheelchair are shown grey shaded. The outdoor part of the route is depicted

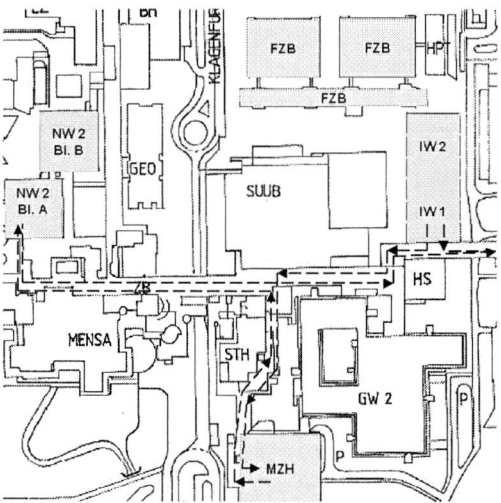

Figure 5.1: *Sketch of the Campus of the Universität Bremen.*

as a dashed line. The exact paths driven indoors are shown in three other figures further below. Including the indoor parts, the route's length amounts to 2,176 m. Traveling along this route with a maximum speed of 84 cm/s takes about 65 min. The average speed was about 56 cm/s. Figure 5.2 shows six photos taken during the "Big Route" experiment. Figure 5.2a depicts a situation in the narrow corridors of the NW2 building. Here, cupboards, lockers, cabinets, and other obstacles line the path. In Fig. 5.2b, the wheelchair has already traveled around 200 m along the main boulevard on its route from the NW2 to the IW building. As shown in the picture, the boulevard is about 30 m wide. In the engineering sciences building IW1, the wheelchair travels along a "gallery"-like balcony through a production hall (see Fig. 5.2c). In addition to fences limiting the path here, the "glass ramp" shown in Fig. 5.2d) turned out to be challenging for the laser range finder. Other problems occurred during the return to the MZH building, shown in Fig. 5.2e), where a lot of bushes line a part of the boulevard that is about 15 m wide. And finally, Fig. 5.2f) depicts the return of the wheelchair to the MZH. As shown, the indoor environments are rather spacious (here: entrance hall of the MZH).

Please note that the "problems" mentioned here do only affect the results of the laser mapping approach used as reference locator. RouteLoc is independent of proximity sensor impressions. As a consequence, the pedestrians strolling across

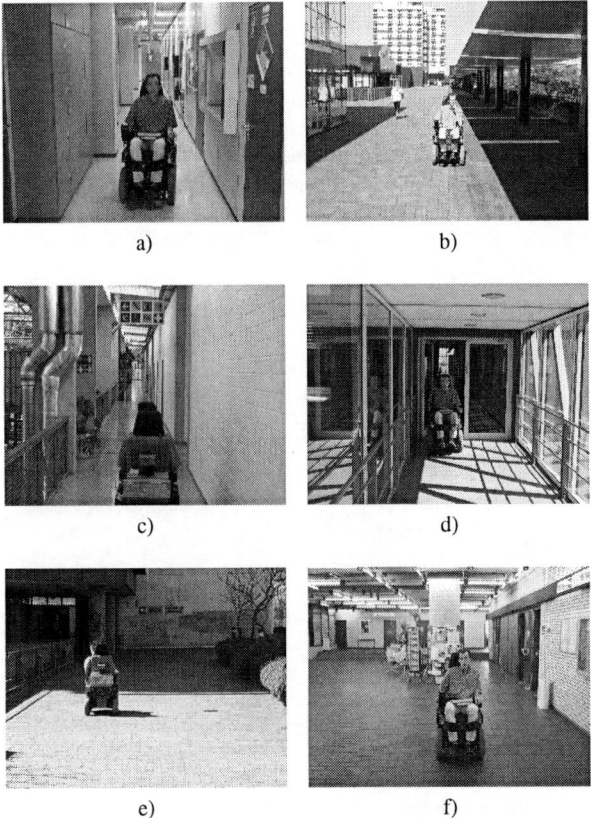

Figure 5.2: *Snapshots from the Campus Route.* a) in the NW2 building, b) from the NW2 building to the IW building, c) in the IW building, d) the "glass ramp", e) returning to the MZH building, f) in the spacious MZH entrance hall. Photos by Markus Eich.

the boulevard and through the entrance halls of the buildings did not matter here. Probably, Choset and Nagatani (2001) refer to situations similar to that depicted in Fig. 5.2e) when they say that the Voronoï-graph based approaches are lost in environments with no or only few features.

While traveling, the wheelchair generated a log file which records one sensor state vector every 32 ms. Such a state vector contains all the information available to the wheelchair: current speed and steering angle, joystick position, current sonar measurements, and complete laser scans. As mentioned, only the locomotion data is used for ROUTELOC. In addition, the algorithm for the incremental route generalization requires the input of two sonar sensors that are mounted at the wheelchair's left and right side. They are used to determine the corridor width. By feeding the log file (192 MB) into the simulation tool SimRobot (Röfer, 1998c), it is possible to test ROUTELOC with *real* data in a simulated world. Note that the simulator works in real-time, i. e. it also delivers the recorded data in 32 ms intervals to the connected software modules, one of which is the self-localization module.

5.1.2 Indoor Environment MZH 3rd floor

Here, the route graph represents the whole third floor of the MZH building, including rooms and unused corridors. In the experiments below, ROUTELOC's performance is evaluated with a sparse route graph (modeling only the real world corridors) and a complete route graph (also representing the offices). For details, see below.

5.2   Evaluation Method

In order to evaluate the performance of an approach for the global self-localization of a mobile robot, a reliable reference is required that delivers the *correct* actual position of the robot. Then, this reference can be used to compare it with the location computed by the new approach, and thus allows assessing the performance of the new method. As mentioned above, ROUTELOC uses a hybrid topological-metric representation of the environment. As a consequence, its position estimates are also "hybrid": a typical estimate would be "the wheelchair is in the segment between decision point **A** and **B** in a distance of, e. g., 256 cm past **A**".

A metric self-localization method is used as a reference. To be able to compare the metric positions determined by the reference locator with the junction/distance pair returned by the new approach, a metric real-world position for all junctions is computed in advance. Thus, it is possible to infer an (x, y, θ) triple from the junction/distance representation that can be compared to the metric position returned by the reference locator.

5.2.1 Scan Matching

The method used as a reference was developed by Röfer (2001a) and is based on earlier work by Kollmann and Röfer (2000). The approach improves the method of Edlinger and Weiß (1995) to build maps from measurements of laser range

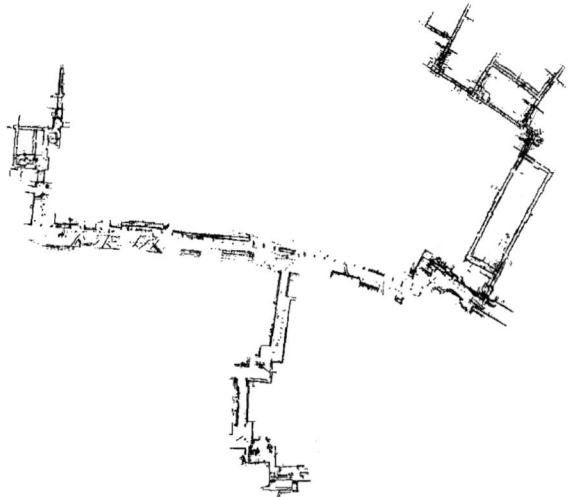

Figure 5.3: *Laser-scan Map of the "Big Route".* The laser-scan map is generated from the data recorded along the route depicted in Fig. 5.1.

finders (laser scanners) using a histogram-based correlation technique to relate the individual scans. Kollmann and Röfer (2000) enhance the original approach by state-of-the-art techniques, namely the use of projection filters (Lu and Milios, 1997), line-segmentation, and multi-resolution matching. The line-segmentation method employs the same approach already used for the route generalization (see Sect. 4.2.1). It runs in linear time with respect to the number of scan points and therefore is faster than other approaches, e.g. the one used by Gutmann and Nebel (1997).

The generation of maps is performed in real-time while the robot moves. An important problem in real-time mapping is consistency (Lu and Milios, 1997), because even mapping by scan-matching accumulates metric errors. They become visible when a loop is closed. Röfer (2001a,b) presents an approach to self-localize and to map in real-time while keeping the generated map consistent. For the latest state of development, please refer to Röfer (2002).

Figure 5.3 shows a map made of laser-scan data that covers the complete "Big Route". As described below, such a map is used by the reference locator to determine the metric position of the wheelchair.

5.3   The Results

This section presents the performance of RouteLoc in the scenarios introduced in the previous section. First, the "Big Route" is examined in detail in Sect. 5.3.1. A series of experiments also comprising a robot kidnapping are described. As a complement, Sect. 5.3.2 shows results from indoor experiments with detailed route graphs.

5.3.1   "The Big Route" – Results

The "Big Route" scenario has already been described in Sect. 5.1.1. Here, the route graph used to represent the environment is presented. In addition, the input data on which RouteLoc has to carry out the self-localization is shown.

Following this, the strengths and also the limitations of RouteLoc are identified. This is done by a step-by-step evaluation of the localization performance on the "Big Route" (see Sect. 5.3.1).

Route Graph and Route Generalization.

The route graph depicted in Fig. 5.4 serves as environment model for the "Big Route" experiments. It covers the part of the campus of the Universität Bremen that is relevant for the route traveled. The route graph consists of 46 graph nodes and 144 junctions. The represented corridors range in length from 4.3 m to 179 m. Please note that only those corridors are represented in the route graph, the wheelchair enters at least once during its trip. This is due to clarity reasons for the figures here. Section 5.3.2 uses a smaller indoor environment to show that the approach also works if the whole environment is represented.

For the performance evaluation, a laser-scan map of the whole route depicted in Fig. 5.1 was generated, using the scan matching method presented by Röfer (2001a). For such a large scene, the laser map deviates from the original layout of the environment in that the relative locations of the buildings are not 100% correct (see Fig. 5.3). Therefore, the route-graph was embedded into the laser-scan map. That way it is possible to compare both localization results on a metric basis while traveling through the route with simultaneously active scan matching and route localization modules. That is the reason why the layout of the route graph depicted in Fig. 5.4 differs from the map shown in Fig. 5.1. While the buildings themselves are correctly represented, the orientations of some route parts are not truly mapped. As an example consider the slight deviation in orientation between segments **rq** and **of**, which are in parallel in reality. A second problem is that the matching of the laser scans fails when the wheelchair leaves the IW building on its way back to the boulevard. Then, another "virtual" exit from the IW building

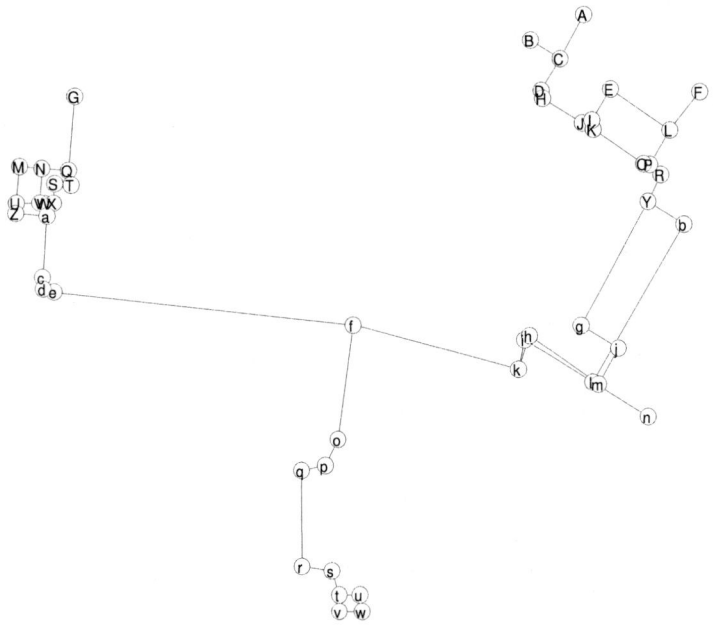

Figure 5.4: *Route Graph of the "Big Route".*

to the boulevard is used by the scan-matching approach. To cover this, the route graph was patched in that region. That's why some segments cross each other in Fig. 5.4.

Please note, that all these "fixes" are only necessary to make feasible the metric evaluation in comparison to the scan-matching approach.

The initial form of the route graph was derived from the ground plans of the buildings (see also Figs. 5.7, 5.10, and 5.13). Then, it was "bent" at the critical points to fit into the laser map depicted in Fig. 5.3.

As situation model, the route generalization approach introduced in Sect. 4.2.1 incrementally provides ROUTELOC with the description visualized in Fig. 5.5. The circles along the route depict the corners found by the generalization algorithm. Those labeled with letters correspond to correctly detected corners, the others are "phantom" corners. The labels correspond to the junction labels used in Fig. 5.4. The arrows indicate the wheelchair's driving direction. The indoor parts of the NW2 and the FZB building are additionally depicted in a "zoomed" version. The

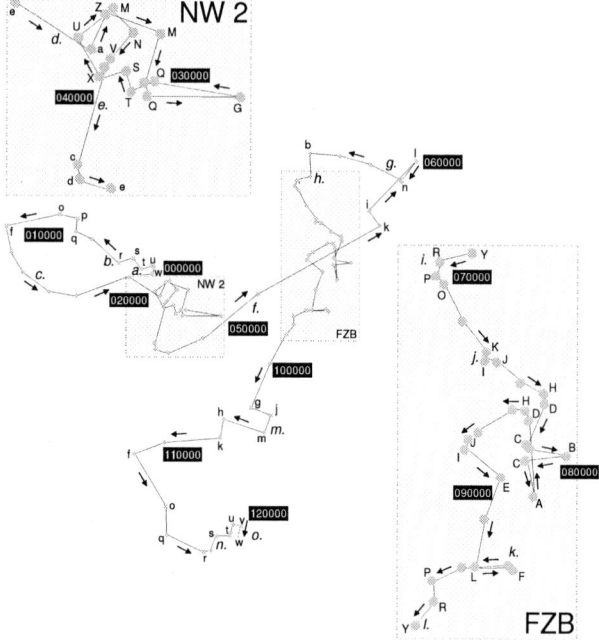

Figure 5.5: *Annotated Route Generalization of the "Big Route" Across the Campus of the Universität Bremen.*

italic letters followed by a dot mark the following interesting places:

a. The route starts in the MZH building, the "frame counter" shows 000000 (slightly to the left in the center of the figure).

b. The wheelchair leaves the MZH through a glass corridor (005200).

c. The "main boulevard" on the way to the NW2 building (013000). Obviously, the data logged by the wheelchair's odometry is not that precise. Whereas distances are measured relatively precisely, angles between segments are often completely wrong: e. g., compare the different generalizations of the "boulevard" here and on the way back (see item f. below).

d. Leaving the boulevard and entering the NW2 building (021000).

e. Leaving the NW2 building heading back to the main boulevard (042000).

f. The main boulevard on the way back (`052000`). This time, this straight "corridor" is generalized almost as a perfect straight line (in comparison to the way there, see item c. above).

g. After having stopped (`059500`) and driven backwards for some meters, the wheelchair enters the IW building at `062000`.

h. In a round corridor, the wheelchair transits from the IW building to the FZB building (`070000`).

i. After the round corridor, the FZB building is entered in a very narrow corridor (`071000`).

j. In the FZB building, three little stairwells have to be passed on a gallery in the first floor. One of them is at this position, `075000`.

k. A turn in the FZB building in frame `092250`.

l. Leaving the FZB building, heading back to the IW building in frame `097750`.

m. Leaving the IW building, heading back to the boulevard in frame `103000`.

n. Re-entering the MZH building after around an hour of travel in frame `117500`.

o. Finish after `121000` frames. Even though start and end point (in the bottom center of the figure) are identical in reality, they differ by 290 m in the recorded data.

Performance.

In order to evaluate the localization performance of ROUTELOC, its estimates are compared with the "real" position delivered by the reference locator along the whole route.

As ROUTELOC represents the environment as edges of a graph, its metric precision is limited. The edges of the route graph are not always centered in the corridors; therefore, deviations perpendicular to a corridor can reach its width, which can be more than 10 m outdoors (e. g. corridor **dc**). There are three reasons for deviations along a corridor: first, they can result from the location at which the current corridor was entered (see Sect. 4.3.4). The bandwidth of possibilities depends on the width of the previous corridor. Second, deviations can be due to odometry errors, because the wheelchair can only correct its position when it changes corridors at a junction. In case of the boulevard ("corridors" **ef**

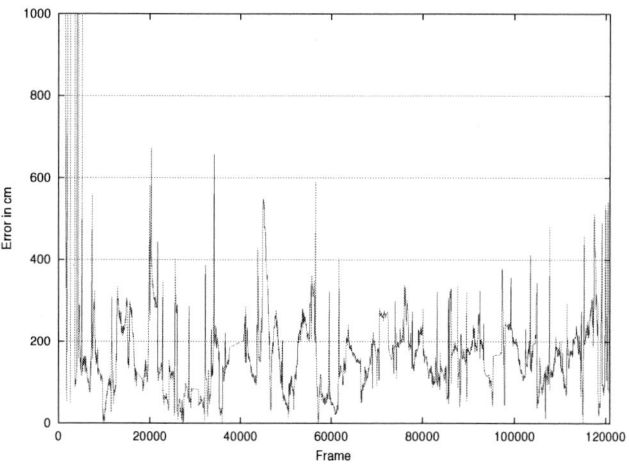

Figure 5.6: *Position Estimate Error (Complete Route).*

and **fk**), the wheelchair has to cover approximately 300 m without the chance of re-localization. Third, deviations can also result from a certain delay before a turn is detected. Such generalization delays are discussed in Sect. 4.3.4 and will be referred to in the specific evaluation further below.

Even though the odometry data turned out to be very bad (see Fig. 4.15), RouteLoc is able to robustly localize the wheelchair. After the initial successful localization, the correct position is maintained over the whole route distance.

Figure 5.6 visualizes the position estimate error of RouteLoc. As will be the case for the following figures (Figs. 5.8-5.16), the difference between the position estimate given by RouteLoc and the one given by the reference locator is used as *y*-value. The horizontal *x*-axis runs through the approximately 121000 frames, the robot encountered during its travel. Remember that each frame is about 32 ms long. That means the *x*-axis does not represent the travel *distance* but the travel time. The wheelchair stopped several times and also had to shunt sometimes. Therefore, the distances along the horizontal axis do not directly correspond to metric distances along the route. The frame numbers can be looked up in Fig. 5.5, where the route generalization recorded during the journey across the campus is visualized and annotated.

The average error during the total route is 471 cm; the median is 159 cm. This is less than a thousandth of the total route length. If only the position estimates after the first "successful" positioning around frame 5200 are considered, the mean error drops to 173 cm (and the median to 158 cm). Please note, that this approach uses as source of information only the generalized route description incrementally acquired during run-time and the a priori known route graph.

The maximum deviation after the initial positioning phase is completed is 673 cm in frame 20329 (at the end of the long main boulevard, cf. Fig. 5.8 below), the smallest error is about 0.1 cm at frame 114997 (on the way back to the MZH building). By chance, the smallest error is outdoors. In general, the position estimates are more precise indoors since the corridors are narrower there.

In 23.1% of all frames (23.5% if considering only those values after the first high confident positioning), the error is below 1 m; it is between 1 m and 2 m in 45.5% (46.2%) of all frames; it is between 2 m and 3 m in 22.3% (22.9%) of the cases; it is between 3 m and 4 m in 5% (5%) of all frames; and it is more than 4 m (but no more than 6.73m) in 4.1% (2.3%) of the cases.

After completing the 2,176 m long route, the deviation of RouteLoc's position estimate from the reference location provided by the scan-matching approach is about 80 cm.

Figure 5.6 shows that it takes a while before the initial uniform distribution adapts in such a way that there is sufficient confidence to pose a reliable hypothesis about the current position of the robot. But if this confidence is once established, the position is correctly tracked. To be precise, the position is not tracked, but absolutely determined, i. e. even when the robot would be deported during runtime to another place, it would be able to re-localize itself (see the discussion of the "kidnapped robot" problem in Sect. 5.3.1). As will be shown below, the regular peaks in Fig. 5.6 mainly result from the generalization delay problem as described in Sect. 4.3.4.

In the following, the results of the whole experiment are discussed in detail. The first part of the route from the start in the MZH building over the main boulevard to the NW2 is shown in Fig. 5.8. The route part in the narrow corridors of the NW 2 building is focused in Fig. 5.11. The way back along the boulevard heading towards the IW building is the subject of Fig. 5.12. The way there through the IW and FZB buildings is visualized in Fig. 5.14, the way back on a different track through the same buildings is shown in Fig. 5.15. Finally, Fig. 5.16 completes the journey with the route part from the IW building over the boulevard back to the MZH building and the starting point.

From the MZH Building to the NW2 Building.   After the wheelchair started its journey in the MZH building, it is RouteLoc's task to localize the robot in the route graph. Due to the initial uniform distribution of the junction's matching qualities, it takes a while before RouteLoc is able to give a correct position estimate

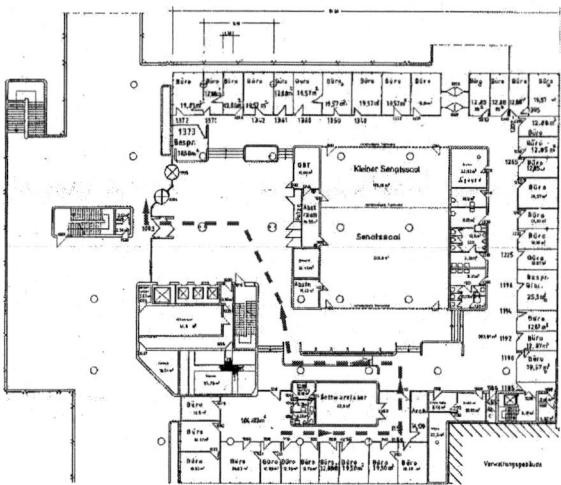

Figure 5.7: *Route (Way There) Through the MZH Building (Ground Plan).* The dashed line is a sketch of the wheelchair's path.

for the first time (approx. in frame 5000, see Fig. 5.8). Even though this period of time (about 2 min.) is rather short in the light of a route length of about 65 min (or 121000 frames), it could be even smaller, if another starting point for the journey was chosen (cf. Fig. 5.17 on page 117). The problem with starting in the MZH building as done here is that traveling in the MZH "looks almost the same" as traveling in the NW2 building. Please compare Fig. 5.7 and Fig. 5.10. The path traveled by the wheelchair in the MZH also fits into the NW2: starting in the upper part of the lower building, the U-shaped part of the wheelchair's path leads to the transition to the upper part of the NW2 building. As a consequence, from frame 1000 to 5200, positions in the MZH building as well as in the NW2 building are favored. The first successful localization is done from frame 3500 to frame 3800. But in the following, there are two situations (frames 4800*ff.* and 5000*ff.*), where the remote twin-position in the NW2 wins — and, consequently, drastically boosts the localization error here. Nevertheless, in frame 5200, RouteLoc gives a position estimate which is less than 2 m away from the reference position determined by the Laser-Matching approach. This is exactly when the wheelchair turns right after leaving the MZH building heading to the boulevard. This right turn is the distinctive feature that allows RouteLoc to doubtlessly estimate the wheelchair's position to be in the MZH building.

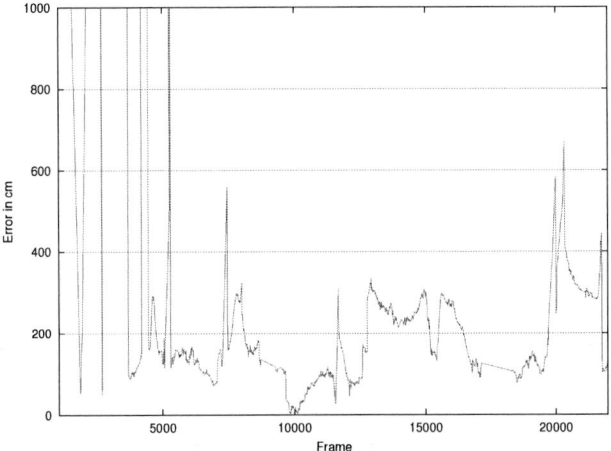

Figure 5.8: *Position Estimate Error (First Route Part from MZH to NW2 Building).*

From that point on, RouteLoc does not lose the position anymore and correctly locates the wheelchair on the campus. Note again that RouteLoc is an absolute approach, i. e. it does not track the movements of the robot, but it carries out a true global positioning. Since RouteLoc uses the probabilistic approach described in the previous sections, a certain inertia occurs when changing corridors. This inertia, the generalization delay discussed in Sect. 4.3.4, leads to some of the peaks that can be observed in Fig. 5.6.

The first significant peak in frame 7200 (see Fig. 5.8) is a good example for the generalization delay problem. Fig. 5.9a-c show the traveled track as recorded by the wheelchair's odometry (upper row) and the incremental route generalization thereof (lower row) around frame 7400. The wheelchair is on its way from the MZH building to the main boulevard. Fig. 5.9a depicts the situation shortly before the wheelchair turns to the right at a 90° right turn (frame 7184, the change from **rq** to **qp** as depicted in the route graph in Fig. 5.4). The generalization (lower row) is a rather appropriate abstract representation of the odometry data. In this situation, the position estimate error is below 2m. As shown in Fig. 5.9b, the wheelchair turns right at the junction. But, due to the reasons discussed in Sect. 4.3.4, this is not acknowledged by the route generalization algorithm at that time. Therefore, the distance between the position estimate (believing to be still in segment **rq**) and the real wheelchair position (already driving in segment **qp**) rapidly in-

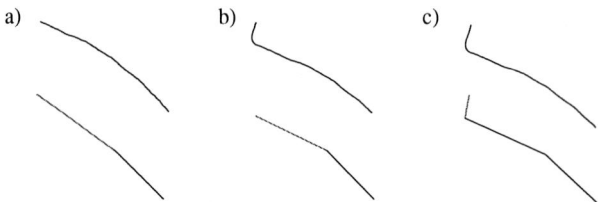

Figure 5.9: *Cause of the Peak Around Frame 7400:* The generalization delay problem as described in 4.3.4. The figure shows real data.

creases. It soon reaches about six meters (in the depicted frame 7460). Figure 5.9c shows the situation which occurred only a good second after the one depicted in Fig. 5.9 (namely in frame 7495). The difference is, that the wheelchair proceeded far enough in the new segment such that the route generalization algorithm detects the corner. As a result, RouteLoc also favors the new junction **rqp** and the position estimate error drops to a "normal" value below two meters.

In frame 10000, the main boulevard is reached. The subsequent deviations are mainly due to lateral errors, i. e. the wheelchair did not travel exactly where the route graph lies (cf. also below, Fig. 5.12). The unusually straight line between frames 17500 and 18500 results from a standstill of the wheelchair (the student who took the photos in Fig. 5.2 had some questions regarding the camera). The line is not horizontal because RouteLoc does not update its position estimate during the standstill. After 1000 frames of "photo stop", the first new position estimate is about 20 cm "better" than the last one before the stop.

The most obvious peak (around frame 20000) is once again due to a generalization delay. After having traveled more than 170 m on the main boulevard, the 90° right turn heading towards the NW2 building takes a while to be detected by the generalization algorithm. But when entering the building in frame 24000, the deviation is less than 2 m.

Interestingly, the error does not increase significantly along the more than 170 m long part of the main boulevard traveled between frames 11000 and 19500.

Within the NW2 Building. The localization task in the NW2 building is more difficult than anywhere else on the route. As Fig. 5.10 shows, the NW2 is a rather symmetric building with narrow corridors that meet in a hall near the entrance of the building. Many paths that can be thought of cannot be unambiguously related to a distinct position at first glance. Due to the probability distribution, RouteLoc acquired so far while traveling, it nevertheless is possible to correctly locate the wheelchair. The route leads the wheelchair through both parts of the NW2 building: in the entrance hall it turns left and drives through narrow corridors until it

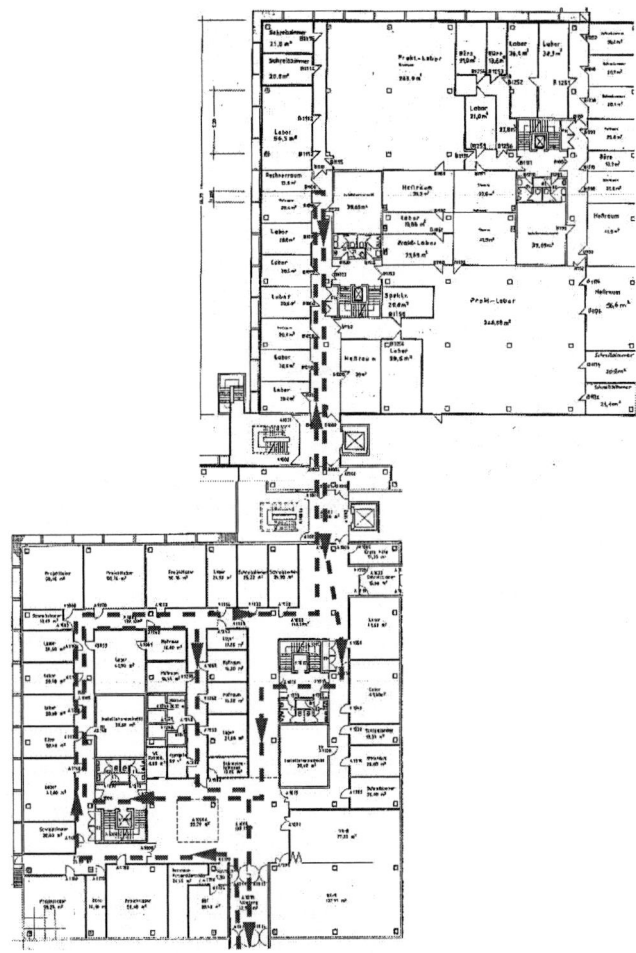

Figure 5.10: *Round Trip Through the NW2 Building.* The dashed line is a sketch of the wheelchair's path.

reaches a piece of open space. There, the wheelchair transits to the B-tower of the NW2 complex. After a while, it turns around and chooses another round trip in

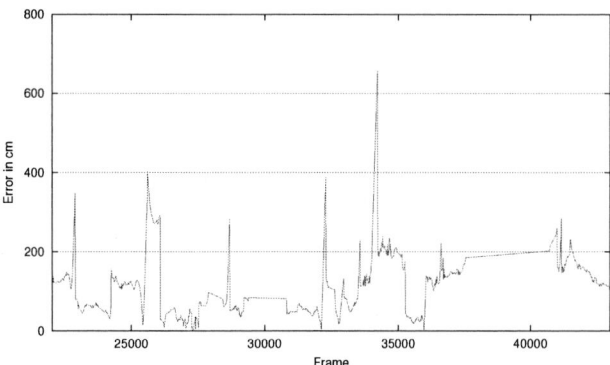

Figure 5.11: *Position Estimate Error (Second Route Part Within NW2 Building).*

the A-tower of the building complex.

Figure 5.11 shows RouteLoc's performance while traveling through the NW2 building. Overall, the deviation between RouteLoc's position estimate and the reference position is very low. There are four interesting situations within the NW2 building:

- The wheelchair passes the same corridor **UM** twice (cf. Fig. 5.4), and both times the error is only around 60 cm. The passages are between frames 23500 and 24300 for the first, and between frames 35300 and 36100 for the second.

- The first turning maneuver on the route traveled up to that point is carried out around frame 28000 (see the corresponding peak in Fig. 5.11).

- Once again, there are two standstill phases: around frame 30000, and a longer one from frame 37800 until frame 41000 in the entrance hall.

- The localization performance is better in the corridors compared to the one within the entrance hall. This wide hall is depicted in the lower part of Fig. 5.10. It is traversed mid-way two times, first horizontally (with respect to the route graph and also the ground plan orientation) between frames 34000 and 35300, and second vertically between frames 37800 and 42000.

From the NW2 Building to the IW+FZB Building Complex. After leaving the NW2 building, the usual generalization delay peak occurs when returning to

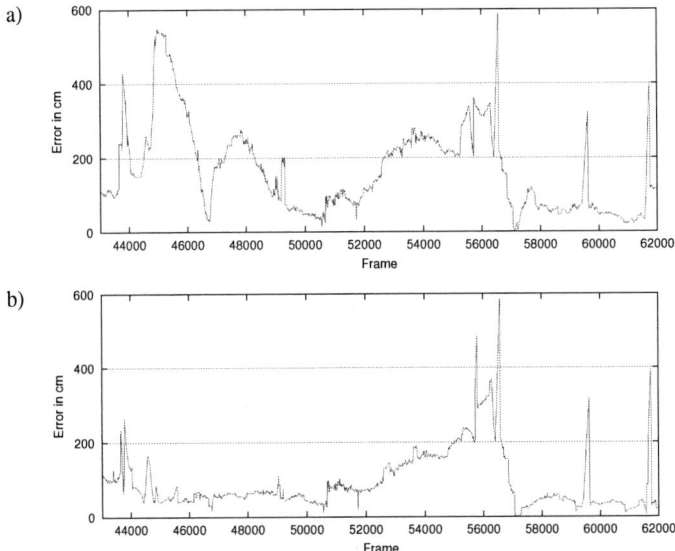

Figure 5.12: *Position Estimate Error (Third Route Part From NW2 to IW Building).* a) The vertical axis shows the position estimate error as usual. b) The vertical axis shows the position estimate error when ignoring lateral deviations by projecting the reference position on the route graph (see text).

the main boulevard in frame 44000 (see Fig. 5.12a). Then, an interesting peak appears in the chart (frames 45000-47000). It results from a maneuver in which an area of pieces of broken glass had to be detoured to the right (also cf. the following paragraph). When returning to the originally intended path along the boulevard (frame 46500), the wheelchair overshoots to the left of the path for a while. Around frame 50500, the deviation between RouteLoc's position estimate and the reference position is less than 20 cm, which is remarkably low for a travel distance of more than 170 m within the long segment **ef** which offers no possibility for recalibration. However, the situation changes during the following 6000 frames where a significant odometry drift can be observed (until frame 56000). The two minor peaks around frame 61000 are caused by generalization delays when turning from segment **ki** to **il** and from **ln** to **nl** (turn), respectively. Note that segment **nl** is driven backwards, i.e. the wheelchair does not turn around. Nevertheless, RouteLoc has to find out that it changed junctions and entered the turn-junction **lnl**.

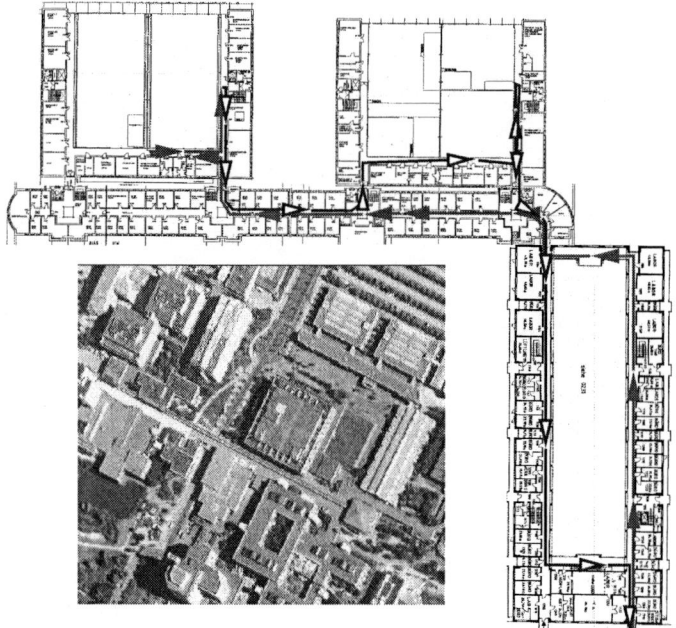

Figure 5.13: *Round Trip Through the IW and FZB Buildings.* To avoid a "white spot", an aerial picture of the campus area is shown in the lower left of the figure. The IW and FZB buildings are in the upper right of the photo. Picture by `www.ausderluft.de`

To make the problem of odometry drift more explicit, Fig. 5.12b shows the same part of the route as the previous Fig. 5.12a. The lateral error was filtered by always projecting the reference position calculated by the scan-matching approach onto the route graph. That way, one ROUTELOC immanent source of error is hidden, namely the deviation resulting from not exactly driving where the route graph is. As a consequence, the detour of the broken glass between frames 44000 and 46500 (cf. Fig. 5.12a) is no longer noticeable. Nevertheless, the odometry drift along the main boulevard remains apparent.

Way There in the IW+FZB Building Complex.   The next building on its route is depicted in Fig. 5.13. Coming from the boulevard, the wheelchair enters the IW building. The dashed line marks the way there and the dotted line marks the way back (also cf. the following paragraph). The IW and FZB buildings stand out for

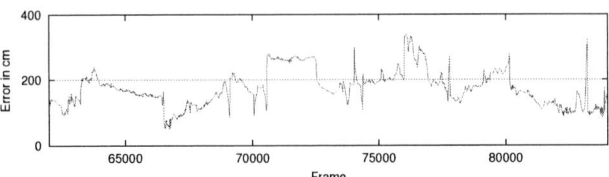

Figure 5.14: *Position Estimate Error (Fourth Route Part Through IW and FZB Buildings, Way There).*

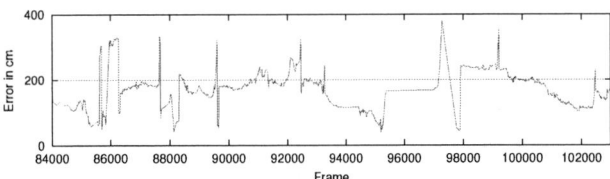

Figure 5.15: *Position Estimate Error (Fifth Route Part Through FZB and IW Buildings, Way Back).*

their narrow corridors, some of which are gallery-like. A "round" corner has to be passed and two turning maneuvers have to be performed. Figure 5.14 shows the typical generalization delays when changing junctions. Overall, the error level is around the average error in this building complex.

Way Back in the IW+FZB Building Complex.   Figure 5.15 shows the way back to the boulevard through the FZB building and the IW building. The driven path is different from the way there. The usual generalization delays can be observed in frames 85750 (junction **CDH**), 86000 (junction **DHJ**), 87750 (junction **HJI**), 89750 (junction **IEL**), 92250 (junction **LFL**), 97750 (junction **PRY**), 99000 (junction **RYg**), and 103000 (junction **Ygj**).

Interestingly, the difficult detection of the turn maneuver in frame 92250 causes no especially high peak. The peculiar shape around frame 96000 is caused by another standstill of the wheelchair in the round corridor that connects the FZB and IW buildings.

From the IW+FZB Building Back to the MZH Building.   Traveling back from the IW to the MZH building takes about eight minutes. The deviation in the position estimate is still less than 3 m except for the short moments of a junction change where the generalization delay happens, as described above. The gener-

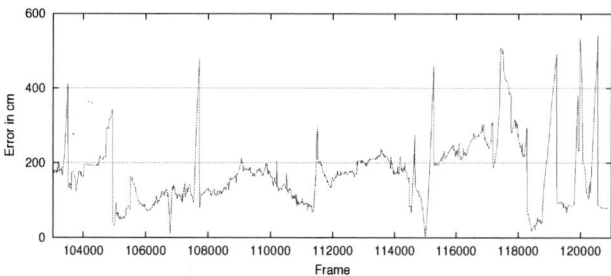

Figure 5.16: *Position Estimate Error (Final Route Part From IW to MZH Building).*

alization delays become especially obvious for the final 4000 frames, where the wheelchair enters the MZH building in frame 117000 (see Fig. 5.16). Since the corresponding route corner is driven on the open boulevard depicted in Fig. 5.2e in front of the MZH entrance, there are no walls that help the route generalization to quickly detect that a new corridor has been entered. As a result, the peak in the chart is relatively high. Within the MZH three more junctions have to be detected. Each new junction once again causes a generalization delay that can be observed between frames 119000 and 120800.

The "Initialization Delay".

As shown in the previous paragraphs, it takes about 5200 frames, i. e. almost three minutes, before RouteLoc is confident about its position estimate for the first time. The time that passes before this first confident position estimate is given, is called "initialization delay". It depends on the specific layout of the environment and the concrete starting position of the robot. If the robot starts at an ambiguous location as shown here (the initial part of the route driven in the MZH also fits in the NW2, as discussed above), it takes rather long to disambiguate the estimation. But please note that "rather long" refers to time or travel distance. When considering the number of "landmarks" (i. e. the number of passed corners in RouteLoc's case), the robot gives a correct position estimate after the fifth corner was detected.

If instead, the robot starts at a more uniquely shaped location, RouteLoc should be able to give a good estimate earlier. To substantiate this claim, consider Fig. 5.17. It shows an experiment, in which the robot starts in the NW2 building with no information about its environment, i. e. the robot starts from scratch with a uniform distribution of probabilities over the junctions. From the new starting position, it follows the same route as in the original experiment. The new starting

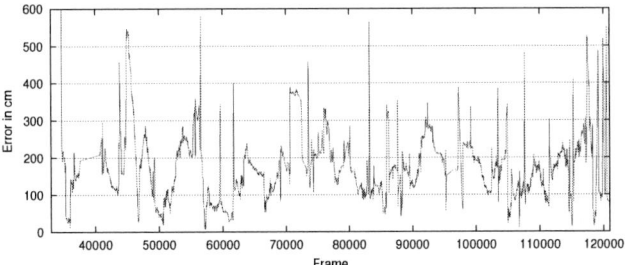

Figure 5.17: The "Initialization Delay".

position is equal to the one the robot occupied in frame 33000 (cf. Fig. 5.5) of the original "Big Route" experiment. It is located in junction **QTS** (cf. Fig. 5.4) shortly after the transition from Block A to Block B of the building (Fig. 5.10). From this new starting position, the robot follows the well-known "Big Route". The difference is that the robot could not acquire any knowledge (i. e. adapt the probability distribution) on its way to the NW2 because it starts there from scratch, this time.

Figure 5.17 shows that the starting situation is rather unique for RouteLoc. After starting its journey there, it takes the robot only 2000 frames (i. e. around one minute) to successfully self-localize. It is already then that there is no more other possible sequence of junctions that matches the generalized route better. As Fig. 5.17 also shows, in this situation it is no disadvantage that the robot did not adapt the probability distribution because the performance for the rest of the route is as convincing as in the "normal" case.

The Kidnapped Robot Problem.

As introduced above, the kidnapped robot problem is considered to be the most challenging task in mobile robot self-localization (e. g. Thrun et al., 2000b). This is due to two reasons: First, a kidnapped robot can only relocalize if it has some means for *absolute* localization. That means that the tracking algorithms do fail here. As will be briefly mentioned below, even some absolute localization approaches fail, because they mutate to tracking algorithms after they are once confident that they correctly localized the robot. As an example, consider the standard Monte Carlo Localization as described by Fox et al. (1999). The second reason is that the robot has to detect that it has been deported and to deliberately unlearn previously acquired knowledge.

Figure 5.18 depicts the error log during a kidnapping experiment. While trav-

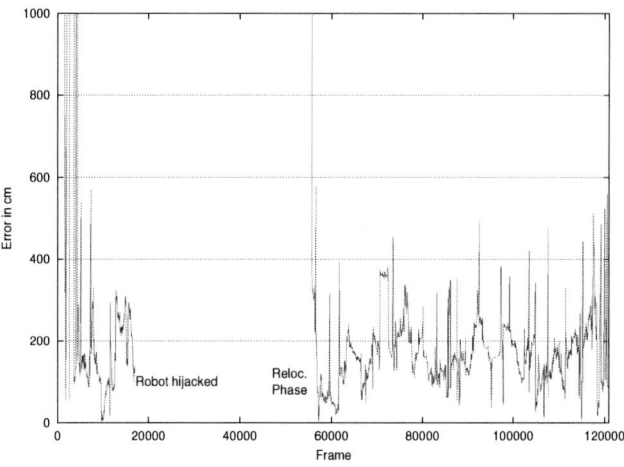

Figure 5.18: Re-Localization after having been Kidnapped.

eling the well-known path of the "Big Route", the robot was "deported" in frame 17000 (on the main boulevard). In fact, the wheelchair was simply turned around on the spot and continued on the Big Route path. In the depicted error chart, the relocation phase of the wheelchair starts in frame 47000. It takes until frame 57000, before RouteLoc is able to estimate the correct position. Please note that this is already after the detection of only one single junction in the route graph. For the rest of the route, the error is similar to the known results from the normal "Big Route" experiment.

5.3.2   Indoor Environment MZH 3rd floor

After having shown the behavior of RouteLoc in a large-scale hybrid indoor and outdoor environment, some details may be clarified in a smaller indoor environment in the third floor of the MZH building.

The first topic addressed is the question how RouteLoc performs in more *complex* environments. Here, "complex" does neither refer to the spatial extent nor to the clutteredness of the environment. Instead, it refers to the degree of detail with which the environment is represented. The route graph in Fig. 5.4 comprises only those corridors, the wheelchair traveled through during the "Big Route" experiments. As can be seen in the corresponding ground plans, at least in the buildings almost all existing corridors are covered by the route. But in general, the route graph should represent the *whole* environment, i. e. also those parts not visited by wheelchair.

The third floor of the MZH building lends itself for this experiment. A sketch of the scene is shown in Fig. 5.20a. This figure already also depicts the route traveled during this experiment. But before, the following subsection describes the route graphs and the route generalization as used in this experiment. Then, ROUTELOC's performance in this environment is evaluated. In the final subsection, a step-by-step analysis of a kidnapped-robot experiment is presented. It was also carried out in the third floor of the MZH building.

Route Graph and Route Generalization.

The third floor of the MZH building is modeled by two different route graph versions:

Sparse Route Graph. The sparse version corresponds to the modeling approach as used for the "Big Route" experiments. Only the corridors used by the robot during its journey are represented in the graph. Similar to the route graph used for the Big Route, this sparse route graph also covers each corridor of the environment (because the robot visited all during its journey). Nonetheless, a set of junctions misses because offices and other rooms are not represented.

Complete Route Graph. The complete version of the route graph comprises a significant number of extra junctions that allow the robot to "enter rooms". Rooms are modeled as about 4 m long dead end "corridors" that depart in 90° angles from the real world corridors.

Both route graph versions are shown in Fig. 5.19. Fig. 5.19a shows the sparse one, Fig. 5.19b depicts the completely modeled one. The labels for the nodes are not shown in Fig. 5.19b, because they are not referred to in the sequel. The route generalization used as input by ROUTELOC is independent of the route graph version. A typical generalization of the route traveled on the third floor of the MZH building is shown in Fig. 5.20b. Please note that the route sketched in Fig. 5.20a is 197 m long. The whole floor extends over an area of 47 m × 49 m.

Performance.

First, the performance of ROUTELOC using the sparse version of the route graph is examined. Then, a comparison with ROUTELOC's behavior when using the complete route graph is made.

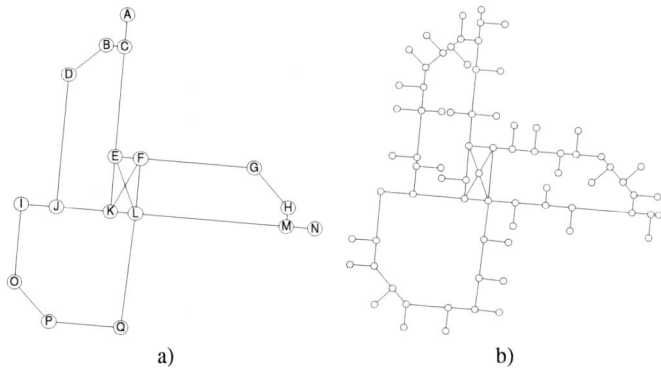

Figure 5.19: *Route Graphs Used for the Indoor Environment on the 3rd Floor of the MZH Building.* a) Only the relevant parts are represented, b) The whole floor is modeled in the route graph.

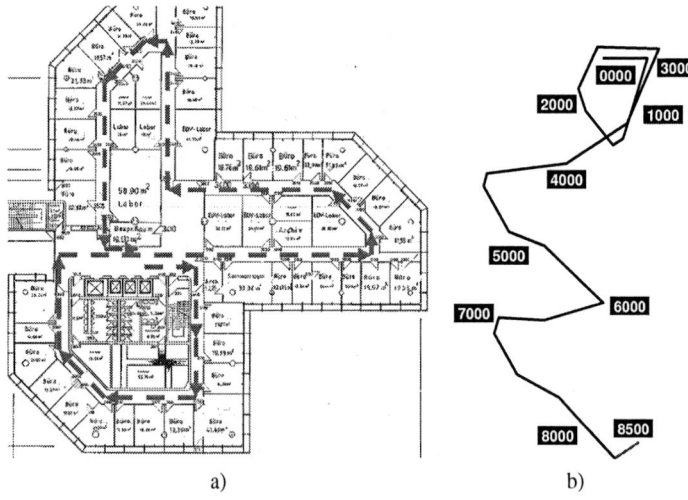

Figure 5.20: *A Sketch and the Generalization of the Traveled Indoor Route.* a) the route sketch is embedded in a ground plan of the third floor of the MZH building; b) The route as perceived by the route generalization algorithm.

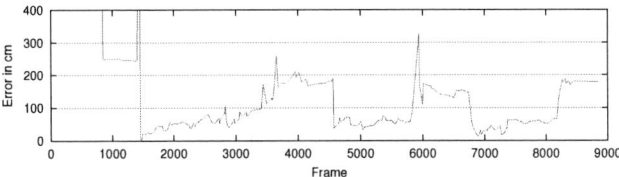

Figure 5.21: RouteLoc's Performance on the 3rd Floor (Sparse Route Graph).

Sparse Route Graph. The performance of RouteLoc using the sparse version of the route graph as shown in Fig. 5.19a is depicted in Fig. 5.21. As shown in the figure, the first successful although not too precise localization is found around frame 900: the position estimate is no more than two and a half meters away from the real robot position. Interestingly, some time later (in frame 1400) the position error drastically increases for a short moment. This is due to the fact that the route generalization detected a corner in the route that is not there in reality (a phantom corner). Nevertheless, a hypothesis exists that this corner might be real. Incidently, there exists a junction in the route graph, namely **QPO**, that would fit convincingly. Therefore, RouteLoc favors junction **QPO** (with the assumption of having started in **LQ**) over the correct position in junction **KLQ** (with the assumption of having started in segment **JK**). A moment later, the route generalization detects the corner that matches junction **LQP**. Whereas the previously favored hypothesis of being in **QPO** leads to a low probability of having conducted a turn to the right, the previous runner-up perfectly matches now: RouteLoc correctly favors junction **LQP**. Such an error does usually only occur during the initial phase of self-localization because the "history" components of the junctions are not that meaningful in the early phases of a route. From frame 1410 on, the position estimate is always "correct", i.e. the candidate junction is correctly identified. The error is in the RouteLoc immanent boundaries, which are discussed in Sect. 4.3.4.

Another interesting observation with respect to Fig. 5.21 is that the performance is sometimes very good (e.g. less than 50 cm error between frames 6800 and 7200), and sometimes "moderate" (e.g. steadily increasing error up to almost 2 m from frame 3000 to 4500). Assuming that the "good" parts are the normal ones, why do the others produce significantly worse results? This depends on the situation: the slow but steady increase between 3000 and 4500 is caused by odometry drift *along* the route. This is the long passage in front of the elevators into the corridor to the right wing. After turning left at the end of the corridor, the error is reduced to about 50 cm. The next increase around frame 5900 is caused by a *lateral* deviation from the route graph due to some detouring maneuver. Finally, the about 180 cm of error in the final part of the route are once again caused by a

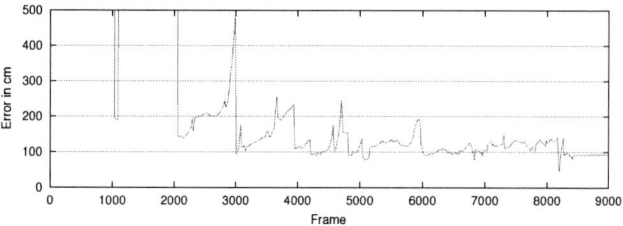

Figure 5.22: ROUTELOC's Performance in the 3rd Floor (Complete Route Graph).

lateral deviation from the route graph.

Complete Route Graph. The comparison with ROUTELOC's performance in a complete version of the route graph shows that a significant increase in junctions does not spoil ROUTELOC's results too much. Figure 5.22 shows the error plot. Two observations can be made: First, it takes about five hundred frames (i. e. around two and a half minutes, or — in this case — 17 m) longer before the first "confident" and also correct position estimate is given; cf. Fig. 5.22 in frame 2000. This is no real surprise: since the environment is modeled by a more complex route graph there exist more ambiguities. Second, apart from the generalization delays around frames 3000 (right turn to the almost open space in front of the elevators, see ground plan in Fig. 5.20a), 4800, and 5900, the position estimate error is around 1 m. This is again in the ROUTELOC immanent boundaries as discussed in Sect. 4.3.4. The increase of the error between frames 3000 and 4000 can easily be explained: The robot travels from the open space in front of the elevators into the long corridor towards the right wing of the building. While traveling, the odometry drift causes the position estimate to deviate more and more from the route graph. Around frame 4000, the route generalization detects that the trajectory has left the "acceptance area" of the currently considered virtual corridor. It therefore generates a new route corner. At that very moment, the position estimate error is cut in half because the position estimate is on the route graph again.

The Kidnapped Robot Problem.

This section is a complement to the kidnapped robot experiment on the "Big Route" in Sect. 5.3.1. Here, the focus is put on the process of relocalization.

Figures 5.23 and 5.24 show a series of snapshots of the ROUTELOC status window during the localization run. Each view depicts the (sparse) route graph with labeled nodes. Furthermore, the current position hypotheses are depicted as tri-

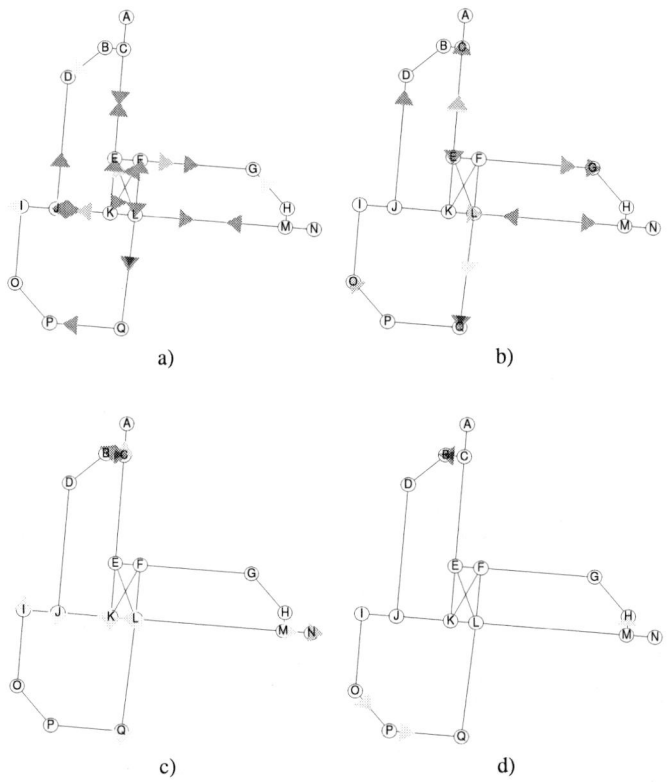

a)   b)

c)   d)

Figure 5.23: *Snapshots During a Kidnapping (I).* a) Frame 1000, i.e. before the kidnapping takes place, b) Frame 3400 (travel and localization resumed), c) Frame 3900, d) Frame 4600.

angles. The intensity of a triangle encodes the confidence in the corresponding position estimate. The candidate junction(s) are marked by a smaller dark triangle within the larger one. In Fig. 5.23a, the situation just *before* the kidnapping is depicted. ROUTELOC already gives a correct position estimate in the junction **KLQ**. Then, the robot is deported by ca. 19 m to junction **OIJ** and proceeds on the route as shown in Fig. 5.20. That means, it follows a straight line within the corridor between decision points **I** and **N**. ROUTELOC has no information about the kidnapping, it assumes the previous estimate to be still true.

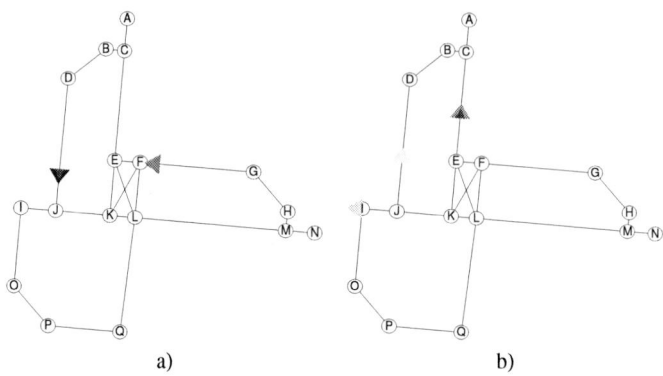

a)   b)

Figure 5.24: *Snapshots During a Kidnapping (II).* a) Frame 5680, b) Frame 6310.

Figure 5.23b shows the situation that arises about 15 seconds after the kidnapped robot resumed traveling. ROUTELOC still favors the junction **KLQ**. It then realizes that its candidate junction cannot host the route anymore, because the traveled route is still straight.

After several seconds, a weak hypothesis proposes at least the correct junction, for the first time after the kidnapping. In Fig. 5.23c, there is a good estimate close to decision point **M**.

The first real success is to be announced about 1600 frames (i. e. 50 sec.) after the robot resumed traveling subsequent to the kidnapping (see Fig. 5.23d). The second best hypothesis is the junction **LMH** in the right wing of the building. This hypothesis covers the real wheelchair position at that time. However, in ROUTELOC's assessment it is only the second best choice. The candidate junction is determined to be in the other wing (junction **ECB**). Note that both wings are not only topologically but also geometrically identical. Please note further that if ROUTELOC were used as self-localization module in a navigation suite, it would be interesting to integrate the planning component. Similar to the approach chosen by Simmons and Koenig (1995), due to the symmetry of the building, the wheelchair would generate the correct action in this situation, even though ROUTELOC's position estimate is more than 35 m off the track (see Fig. 5.25, frame 4600).

For the following 30 seconds, both hypotheses are maintained, whilst the wrong one is still favored (see Fig. 5.24a). But the disambiguation is not far: the wrong, "candidate" hypothesis is going to transit from junction **BDJ** to junction **DJI**, while the correct, "runner-up" hypothesis transits from junction **HGF** via junction **GFE** to **FEC**.

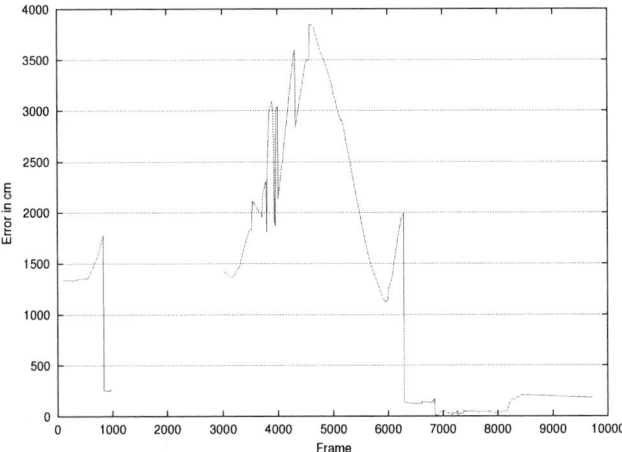

Figure 5.25: *Plot of the Positioning Error During a Kidnapping Experiment.* The gap between frames 1000 and 3000 marks the kidnapping period.

In Fig. 5.24b, the erroneously favored previous hypothesis got stuck in junction **DJI**, which offers no straight extension past decision point **I**. Therefore, the likelihood of this hypothesis drops. Instead, the previously (also erroneously) ignored hypothesis now wins because the junction **FEC** offers some more distance to travel. As shown in Fig. 5.25, the position error is below 150 cm here. For the rest of the route the correct hypothesis is maintained.

5.4   Assessment

Self-Localization of mobile robots in large-scale environments can be efficiently realized if a hybrid representation of the environment is used. RouteLoc matches an incremental generalization of the traveled route with an integrated topological-metric map, the *route graph*. Real-world experiments at the Universität Bremen have shown the robustness and efficiency of RouteLoc.

5.4.1   Pros and Cons of RouteLoc

As the discussion in the previous sections has shown, RouteLoc already works well in the presented version. Nevertheless, it should be regarded as a basic method for absolute self-localization that can be extended on demand.

The strengths of ROUTELOC definitely are its simple needs with respect to sensor equipment, memory and computing time. Since it is only based on the incremental generalization of the route traveled so far, it also works in crowded or completely open environments where sensor based approaches had their difficulties. This is advantageous, as Choset and Nagatani (2001) put it. ROUTELOC is an absolute self-localization approach. On the one hand, this means it is able to find its location from scratch having no need to obtain any initial information except from the route graph. On the other hand, this means ROUTELOC is able to solve the kidnapped robot problem. This is even more difficult than absolute localization because already acquired knowledge has to be unlearned deliberately. The precision of the position estimates is promising because the error is bounded and small enough to solve challenging large-scale navigation tasks. The position estimates themselves are given in a "human-compatible" fashion in the form of "You are 10 meters past your office in the corridor to the conference room". Due to its very basic data structure, ROUTELOC lends itself to a whole variety of possible extensions, as discussed below.

To summarize, ROUTELOC fulfills all the requirements introduced in Sect. 3.3. However, there are also limitations to the algorithm some of which might be resolved in the future, whereas others are method-immanent.

The minimal requirements with respect to the sensor equipment are advantageous for several reasons, but they are also the cause for some lengthy waiting phases before the initial localization succeeds. In this context, Kuipers and Beeson (2002) discuss the trade-off between little sensor usage (which fosters the problem of perceptual aliasing) and complex sensor usage (which fosters the problem of image variability). A subtle disadvantage of ROUTELOC is that its performance depends on the performance of the underlying generalization algorithm. Interestingly, one can also put it the other way round: ROUTELOC profits from being independent from the route generalization algorithm such that various approaches can be tried. Because of the compact representation of the environment as a graph structure, there are system-immanent errors due to the fact that all position estimates are positions on the graph. But especially in wide corridors, it is not necessarily given that the wheelchair follows the graph exactly. A final drawback that might be overcome in the future is that, so far, the route graph must be known to ROUTELOC a priori. A *SLAM* extension has not yet been implemented, see also Sec. 5.4.3.

5.4.2 ROUTELOC in Comparison With Other Approaches

This section compares the new ROUTELOC algorithm for mobile robot self-localization to some other prominent approaches.

"Markov or not?"

A number of prominent self-localization algorithms use the Markov localization approach, some of them with topological representations of the environment (Nourbakhsh et al., 1995; Simmons and Koenig, 1995; Tomatis et al., 2001b), others with metric maps (Burgard et al., 1997; Fox et al., 1999; Thrun et al., 2000a). In the robotics community, this is referred to as "Markov localization" if the algorithm somehow exploits the so-called *Markov assumption* (e. g., see Russell and Norvig, 1995): it states that the outcome of a state transition may only depend on the current state and the chosen action. Explicitly, the outcome does not depend on previous states or actions.

ROUTELOC is not a pure Markov localization: while the matching and propagation process as presented in Sect. 4.3 satisfies the Markov assumption, the necessary handling of the missed junctions and phantom corners violates it.

Comparison Between ROUTELOC and Three Prominent Approaches.

Apart from the "Markov or not?" question, ROUTELOC differs from other localization approaches with respect to some aspects that are gathered in Tab. 5.1. The topological-metric approaches used for the office delivery robots DERVISH by Nourbakhsh et al. (1995) and XAVIER by Simmons and Koenig (1995), and the metric Mixture-MCL algorithm (an improved version of the common Monte Carlo Localization approaches) by Thrun et al. (2000b) are discussed here as reference algorithms.

DERVISH by Nourbakhsh et al. (1995). Nourbakhsh et al. (1995) present an algorithm for the navigation of a mobile robot in an office environment. Their robot DERVISH won the Office Delivery event of the 1994 Robot Competition and Exhibition, held as part of the 13th National Conference on Artificial Intelligence. The interesting component for this review is the so-called "Hallway Navigator" which is responsible for localizing the robot in corridors. The environment is minimalistically represented as a connectivity map that contains no metric information except from approximate corridor and doorway widths.

DERVISH uses a set of sonar sensors to detect features along its routes. It is able to distinguish four different features on either side of the robot: wall, open door, closed door, corridor. The orientation of the robot is kept constant, its odometry is only used for centering within corridors.

DERVISH solves the localization task by maintaining a so-called state-set. Each state represents an area in the real world that encompasses a decision point or a corridor between two decision points, i.e. nodes and edges, respectively, in the topological map. Furthermore, each state is associated with a certainty factor that

Table 5.1: Comparison Between RouteLoc and Three Other Approaches

Model	RouteLoc	Nourbakhsh *et al.*	Simmons & Koenig	Thrun *et al.*
Model	Topological-metric map	State set / topological map	Markov states, hybrid topological-metric map	Particle filter, metric
Sensor input	odometry (+ 2 sonars for generalization)	sonars (+ odometry for corridor centering)	odometry + sonars	odometry + camera or laser range finder
Scenario	campus (in & outdoor)	office (indoor)	office (indoor)	museum (indoor)
Markov	no	yes	yes	yes
Complexity	144 junctions + 102 turn-junctions for 46 nodes and 100 edges, depends on number of decision points	One state per node (decision point) or per edge (corridor)	3348 Markov states for 95 nodes and 180 edges, depends on metric extent of environment	About 1000 samples for an indoor environment, number of samples adaptable (anytime)
Memory	low	low	low	huge
Precision	Position estimate given by junction and metric offset	Corridor resolution, no metric information	Markov states provide resolution of 1 m (translational), 90° (rotational)	Samples indicate position, only small errors

represents the probability that the robot is located in the corresponding area. The localization resolution of the approach is rather limited, since its world model restricts the precision to the extent of the states used. As each state either represents a corridor or a decision point, no information at all is available about the respective position within a corridor.

Nourbakhsh et al. (1995) use an assumptive planning approach: the "state-set progression" technique. After each new sensor percept, the state-set is updated. Then, the state with the highest certainty factor is assumed to correspond to the current position of the robot. Accordingly, the robot proceeds on the intended path to its goal until the goal is reached or the most likely state is not on the path. Even though DERVISH was the only robot that solved the final task, Nourbakhsh et al. recommend to incorporate the available metric information in order to avoid longer phases of "blind flight".

XAVIER by Simmons and Koenig (1995). Simmons and Koenig (1995) improve the results presented by Nourbakhsh et al. (1995) for use in their XAVIER robot. They use partially observable Markov models to track XAVIER's location, i.e. their approach is *not* able to absolutely localize the robot.

The information stored in the Markov model allows to plan and monitor the execution of actions. The approach successively adapts a probability distribution over the possible locations of the robot in the environment. In the presented experiments, their robot is about 90% sure about its initial position. The remaining 10% of the probability mass is spread in the direct vicinity of the robot.

The method is able to handle various kinds of uncertainty: imprecise knowledge of the environment, actuator uncertainty, and sensor noise. The improvement over Nourbakhsh et al. (1995) is that Simmons and Koenig (1995) use a hybrid topological-metric map of the environment. As a consequence, the position estimate is far more precise, because it refers to Markov states which represent one square meter and 90° of orientation each.

MIXTURE-MCL by Thrun et al. (2000b). Thrun et al. (2000b) develop the so-called Mixture-Monte Carlo Localization (MCL) algorithm. It combines the strengths of the standard MCL approach and a derivative of it, the so-called "dual MCL".

Standard MCL spreads a set of samples in the environment and adapts their importance according to sensor perception. In addition, the samples "move" through the environment in accordance to the locomotion of the robot. Over time, the samples gather close to the current position of the robot. They allow to make a position estimate by calculating a kind of "center of gravity" of the sample cluster.

The dual MCL approach tackles the problem the other way round. It guesses states corresponding to the most recent sensor perception, and adjusts the importance factor. The dual MCL is less precise than the standard MCL algorithm, but its position estimate error is monotonic in sensor noise, which is quite interesting.

The integration idea is that the weaknesses of each of the two algorithms can be compensated by the other one. Standard MCL has problems in situations where the perceptual likelihood is too peaked. The dual MCL approach completely fails if confronted with erroneous sensor readings. Indeed, their integration to the Mixture-MCL algorithm delivers the expected promising results.

In real world experiments, Thrun et al. (2000b) find that the Mixture-MCL clearly outperforms other MCL or Markov localization approaches. They also present a kidnapped robot experiment, where the (simulated) robot is deported to a remote position. It remains unclear, how far away from the previous position the "remote" position is. But the first position estimate after the kidnapping is only about 3 m worse than the one before. Nevertheless, it takes a while ("50 time steps") before Mixture-MCL is able to relocalize the robot, albeit less precise than before the kidnapping. As a side remark, note that the standard MCL is *not* able to relocalize at all after a kidnapping took place, due to a lack of samples at the remote position.

5.4.3   Hints for Future Work

This section briefly sketches some ideas how future work could improve on the results presented in this thesis and how the application scenario could be broadened.

Integrating Features

A disambiguation of situations and, as a result, a time reduction for the initial localization can be obtained if the route generalization and the route graph are augmented by feature vectors. Such feature vectors have to be added to the junctions in the graph as well as to the corners in the route generalization. Then, a simple extension of the matching quality definitions introduced in Sect. 4.3.2 allows to improve the confidence about the position estimate in certain situations. The standard solution would be to use additional sensors such as a video camera to obtain the features. But it is also very challenging to exploit the human driver as a source of information here: By introducing an adequate means of dialog about environment characteristics to the system, this information could also be used as a feature. The most challenging problem then is to solve the propagation of these features through the route graph.

Building Route Graphs

The algorithm should be extended such that self-localizing becomes possible even in a priori unknown environments. For this purpose, the robot has to build the route graph from scratch during runtime. Subsequently, it has to solve the problem of place integration (Werner et al., 2000): it has to find out whether its current position is already represented in the route graph, or whether it is located in a corridor that is unknown so far. Moratz et al. (2002) propose the so-called "Ternary Point Configuration Calculus" (TPCC) that allows to detect cycles traveled in a route graph. This approach could be helpful to extend RouteLoc to a SLAM-technique.

Please note that the map building component can no longer be guaranteed to run in real time. As mentioned above, the computational complexity of RouteLoc depends on the number of junctions. If the route graph is build from scratch while the robot moves, this number can grow arbitrarily — at least theoretically. Nevertheless, from a practical point of view RouteLoc's complexity is so low that even the huge amount of some ten thousands of junctions that might be necessary to model a complete city (for Berlin, see Behrens (2002)) could be maintained in "effective real-time", i. e. the theoretical loss of the real-time property would not be relevant in practical use.

Integration of Metric Maps

RouteLoc is well suited to become part of a set of localization modules that work together to conduct the self-localization of the mobile robot. For instance, RouteLoc could be combined with a metric approach that works well in small-scale environments but is not useful in large-scale ones. Another idea is to annotate the route graph with certain attributes that indicate which kind of self-localization approach should be used if the robot were at the assumed position.

Probability Distribution per Junction

RouteLoc's flexibility could be increased by using a probability distribution also for the position within a junction.

Avoiding Generalization Delays

As discussed throughout the results section, the position estimates given by RouteLoc are surprisingly precise taking into account the rudimentary sensor use. Nevertheless, the results are slightly spoiled by the regular error peaks that happen due to the phenomenon of a generalization delay as discussed in Sect. 4.3.4. Two approaches lend itself to improving RouteLoc's results here:

Remove the cause: By improving the capability of the route generalization to detect the transit from one corridor to the next, the problem of generalization delays would be solved at once. For instance, the orientation of the robot could be taken into account when deciding whether a new route segment has been entered.

Fight the symptoms: By introducing a certain inertia into ROUTELOC, some of the peaks caused by a generalization delay could be filtered. As a result, a position estimate that significantly differs from those of the recent history would be ignored for a while.

"Beaming Junctions"

ROUTELOC should be able to cover multi-storey navigation by the incorporation of a new kind of junction, the so-called "beaming junctions." By completely connecting so-called "socket junctions" in each floor with those of the other floors, the transition between floors could be modeled. Please refer to Werner et al. (2000) to find a general discussion on the integration of different layers of route graphs by so-called *transitions*.

"Car-like" Navigation

It would be interesting to evaluate ROUTELOC's performance in "larger-scale" environments such as whole cities. The wheelchair is already able to travel around for several hours, thus experiments in a part of a city should be feasible. But there has to be either some a priori known route graph representing this area, or the map building skill mentioned above has to be realized first. A further requirement is that the route generalization algorithm should only use proprioception to decide whether or not a new route segment has been entered.

Part III

Mode Confusion in Human-Robot Interaction

134

6

There's an old story about the person who wished his computer were as easy to use as his telephone. That wish has come true, since I no longer know how to use my telephone.

Bjarne Stroustrup

Introduction and Motivation

This and the other chapters in Part III are the result of close coopera-tion with Jan Bredereke, which significantly enhanced the author's earlier work on the topic (Lankenau, 2001; Lankenau and Röfer, 2000). A short version of this chapter was published as Bredereke and Lankenau (2002).

Automation surprises are ubiquitous in today's highly engineered world. We are confronted with *mode confusions* in many everyday situations: When our cord-less phone rings while it is located in its cradle, we answer the call by just lifting the handset — and inadvertently cut it when we press the "receiver button" as usual with the intention to start speaking. We get annoyed whenever we overwrite some text in the word processor because we had hit the "Ins"-key earlier (and thereby left the insert mode!) without noticing. The American Federal Aviation Administration (FAA) considers mode confusion to be a significant safety con-cern in modern aircraft. So, it's all around — but what exactly is a mode, what defines a mode confusion situation and how can we detect and avoid automation surprises?

As long as no rigorous definition is provided, a *mode confusion* should be regarded as one kind of an automation surprise. It refers to a situation in which a technical system can behave differently from the user's expectation. Whereas mode confusions in typical human-computer interactions, such as the word pro-cessor example mentioned above, are "only" annoying, they become dangerous if safety-critical systems are considered. For instance, consider the crash of an Airbus A320 near Strasbourg, France, in 1992. Probably due to heavy work-load because of a last-minute path correction demanded by the Strasbourg airport tower, the pilots confused the "VERTICAL-SPEED" and the "FLIGHT-PATH-ANGLE" modes of descent. As a result, the Air Inter flight descended far too steeply, crashed, and 87 people were killed.

Today, many safety-critical systems are so-called embedded shared-control systems. These are interdependently controlled by an automation component and a user. Examples are modern aircraft, automobiles, but also intelligent wheelchairs. This work focuses on such shared-control systems. The entirety of technical components are called *technical system* and the human operator is referred to as *user*. Note that it is necessary to take a black-box stand, i. e. only the behavior of the technical system observable at its interfaces can be considered: since the user's problems are to be solved, the designer has to take his or her point of view, which does not allow access to internal information of the system.

As Rushby (2001) points out, in cognitive science it is generally agreed upon that humans use so-called *mental models* when they interact with technical systems in general, and with automated systems in particular. There are at least two completely different interpretations of the notion "mental model" in the pertinent literature. Therefore, it is important to clarify that the one introduced by Norman (1983) is referred to here: A mental model represents the user's knowledge about a technical system, it consists of a naïve theory of the system's behavior. Sasse (1997) surveys mental models in human-computer interaction in detail.

According to Rushby (2001), an explicit description of a mental model can be derived, e. g. in the form of a state machine representation, from training material, from user interviews, or from user observation. Cañas et al. (2001) survey work on this and show in three experiments with 140 participants how exposing users to different knowledge elicitation tasks allows to figure out their mental models.

Javaux (1998, 2002) works on finding so-called "minimal safe mental models" for specific human-machine interaction tasks. A minimal safe mental model describes the knowledge the user must at least have to be able to interact with the system as required by the safety specification.

This chapter clarifies the notions of "mode" and "mode confusion". As far as the author is aware, there is no publication to date that defines these notions rigorously. Section 6.1 surveys the pertinent state of the art. Section 6.2 introduces a case study, which later serves as a running example. Section 7.1 and 7.2 present a suitable system modeling approach and clarify different world views, which make it possible to present rigorous definitions in Ch. 8. Chapter 9 works out the value of such definitions, which comprises a foundation for the automated detection of mode confusion problems and a classification of mode confusion problems by cause, which in turn leads to recommendations for avoiding mode confusion problems. The approach is then applied to a case study in Ch. 10, which also includes an automated detection of mode confusion problems. A summary and ideas for future work conclude this thesis part.

6.1 Survey of Related Work

The pertinent state of the art is briefly recapitulated here. Since the early 1990s, a number of research groups from the human factors community, in particular the aviation psychology community, have been working on mode confusions in shared-control systems. Recently, people from the computer science community, especially the formal methods community, have also become interested in this topic. There are some promising results with respect to tool supported detection of mode confusion problems (see below), but it remains surprisingly unclear what a mode actually is.

6.1.1 Definitions of Mode and Mode Confusion

While some relevant publications give no (Butler et al., 1998; Crow et al., 2000) or only an implicit definition (Hourizi and Johnson, 2001a; Rushby, 2000) of the notions "mode" and "mode confusion", there are others that present an explicit informal definition (Buth, 2002; Degani et al., 1999; Leveson et al., 1997; Sarter and Woods, 1995).

Lüttgen and Carreño (1999) use the notion "mode" synonymously to "state", and explicitly define that every user input causes a mode change.

Doherty (1998) presents a formal framework for interactive systems and also gives an informal definition of "mode error". He defines modes as partitions of the state-space and uses them to define so-called "user relevant abstractions" based on Hoare's trace model (Hoare, 1985).

Thimbleby (1990) develops his "mode" definition via some intermediate stages from a generic and informal one ("a mode is a variable information in the computer system affecting the meaning of what the user sees and does", (Thimbleby, 1990, p. 228)) to a formal one. In doing so, he focuses his scope to the pure two-agent interaction between the human and the machine. He does not consider the physical environment. The latter is a third agent relevant in shared-control systems. As a result, he defines a mode to be a "mathematical function mapping commands to their meanings within the system" (Thimbleby, 1990, p. 255). Thimbleby does not deal with the mode confusion problem. He therefore does not provide a rigorous definition of the notion "mode confusion".

Wright and colleagues give explicit but example driven definitions of the notions "error of omission" and "error of commission" by using CSP (Hoare, 1985; Roscoe, 1997) to specify user tasks (Wright et al., 1994).

6.1.2 Modeling and Tool Support

Interestingly, the kind of modeling often seems to be influenced significantly by the tool that is meant to perform the final analysis.

Degani and colleagues use State Charts to separately model the technical system and the user's mental model (Degani and Heymann, 2000). Then, they build the composition of both models and search for certain states (so-called "illegal" and "blocking" states) which indicate mode confusion potential.

Butler et al. (1998) use the theorem prover PVS to examine the flight guidance system of a civil aircraft for mode confusion situations. They do not consider the mental model of the pilot as an independent entity in their analysis.

Leveson and her group specify the black-box behavior of the system in the language SpecTRM-RL that is both well readable by humans and processible by computers (Leveson et al., 1997; Rodriguez et al., 2000; Zimmermann et al., 2000). In Leveson et al. (1997), they give a categorization of different kinds of modes and a classification of mode confusion situations.

Thimbleby (see above) uses the so-called PIE modeling approach (Thimbleby, 1990) that describes human-machine interaction by specifying a sequence of user commands, the Program. Such a program is interpreted by the technical system by an Interpretation function and causes some Effect. PIE models are also readable by humans and processible by computers.

Sage and Johnson (2002) describe a rapid prototyping approach for an air traffic control system. They are able to verify safety properties based on a LOTOS system specification and claim that their method can support the operator directed design process proposed in Vakil and Hansman, Jr. (2002) (see below). Nonetheless, they do not specify the mental model of the user.

Rushby and his colleagues employ the Murφ model-checking tool (Crow et al., 2000; Rushby, 2000; Rushby et al., 1999). Technical system and mental model are coded together as a single set of so-called Murφ rules. In each step, all rules are "fired" of which the condition is true; i.e. some manipulation of global state variables is performed. Furthermore, a set of invariants is checked. The mode confusion situations are detected with these invariants.

Lüttgen and Carreño (1999) examine the three state-exploration tools Murφ, SMV, and Spin with respect to their suitability in the search for mode confusion potential. They find that each tool has its advantages but also its drawbacks: Spin supports the designer in finding the sources of mode confusion situations through the animation of diagnostic information. SMV bears the advantage that it integrates temporal logics, while Murφ provides the best specification language.

Buth (2002) and Lankenau (2001) clearly separate the technical system and the user's mental model in their CSP specification of the well-known MD-88-"kill-the-capture" scenario and in a service-robotics example, respectively. The support of this clear separation is one reason why Buth's comparison between the Murϕ tool and the CSP tool FDR favors the latter (Buth, 2002, pages 209-211).

Meanwhile, Rushby (2002) also acknowledges this need to separate both entities. For mode confusion detection, he affirms the advantages of model-checking tools for process algebras such as FDR over tools such as Murϕ. A conformance relation between two descriptions has to be checked. The concepts of refinement and abstraction are required for this. They are provided directly by FDR.

6.1.3 Case Studies

Almost all publications refer to the aviation domain when examining a case study: an MD-88 (Buth, 2002; Leveson et al., 1997; Miller and Potts, 1999; Palmer, 1995; Rushby et al., 1999; Sarter and Woods, 1995), an Airbus A320 (Crow et al., 2000; Hourizi and Johnson, 2001a), or a Boeing 737 (Rushby, 2000). For a non-aviation case study, refer to Thimbleby's running (pedagogical) example, a calculator (Thimbleby, 1990).

6.1.4 Recommendations and Critique

Rushby (2001) proposes a procedure to develop automated systems which pays attention to the mode confusion problem. The main part of his method is the integration and iteration of a model-checking based consistency check along with the mental model reduction process introduced by (Crow et al., 2000; Javaux, 1998).

Vakil and Hansman, Jr. (2002) recommend three approaches to reduce mode confusion potential in modern aircraft: pilot training, enhanced feedback via an improved interface, and, most substantially, a new design process (ODP, for operator directed design process) for future aircraft developments. ODP aims at reducing the complexity of the pilot's task, which may involve a reduction of functionality.

Hourizi and Johnson (2001a,b) generally doubt that avoiding mode confusions alone helps to reduce the number of plane crashes caused by automation surprises. They claim that the underlying problem is not mode confusion but what they call a "knowledge gap", i. e. the user's insufficient perception prevents him or her from tracking the system's mode.

6.2 Case Study Wheelchair

The case study presented here has a service robotics background: it deals with the cooperative obstacle avoidance behavior of the Bremen Autonomous Wheelchair "Rolland".

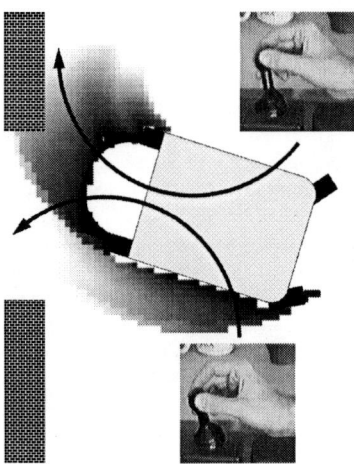

Figure 6.1: *Deciding on Which Side the User Wants the Obstacle to be Passed.*

As mentioned in Ch. 2, Rolland is a shared-control service robot that realizes intelligent and safe transport for handicapped and elderly people. In contrast to other service robots, Rolland is jointly controlled by its user and by a software module. Depending on the active operation mode, either the user or the automation is in charge of driving the wheelchair. Conflict situations, often caused by mode confusions, arise if the commands issued by the two control instances contradict each other. The system architecture and other details about Rolland can be found in Ch. 2. Here, the focus is on the obstacle avoidance skill.

Obstacle Avoidance Skill.

Obstacle avoidance for shared-control rehabilitation robots does not only mean stopping in front of obstacles. In this scenario, two requirements have to be satisfied. Firstly, the handicapped user is to be supported when braking or detouring around objects by the automation such that a smooth and comfortable driving behavior is realized. Secondly, the user should not be surprised, i. e. whatever the automation decides to do, it has to be consistent with the user's expectation. The obstacle avoidance skill realizes an intelligent shared-control behavior by projecting the anticipated path of the wheelchair into a local obstacle occupancy grid map. Figure 6.1 shows a situation in which the wheelchair is supposed to pass through a doorway. The right doorpost is a relevant obstacle since it is on the current path of the vehicle. The distance before collision is visualized for positions on this path by the grey-shaded area: the darker the sooner some part of the wheel-chair will reach the corresponding position. The joystick command shown on the photo in the upper right corner of the figure indicates a narrow right curve. Since the corresponding projected path (upper arrow) points to the right of the doorpost, this command is interpreted as "do not pass through the doorway". The lower photo shows a joystick command indicating a left curve. Since the corresponding projected path (lower arrow) points to the left of the doorpost (i. e. through the doorway), this command is interpreted as "pass through the doorway".

Thus, depending on the side on which the projected path indicated with the joystick passes the obstacle, the algorithm decides how to change speed and steer-ing angle in order to serve the user best. If, instead, the driver directly steers toward an obstacle, the algorithm infers that he or she wants to approach the object and does not alter the steering angle. As a result, obstacles are smoothly detoured if desired, but they can be directly approached if need be. If the automation realizes that the projected path of the vehicle happens to be free after an avoidance ma-neuver, it again accelerates up to the speed indicated by the user via the joystick.

The transition to the obstacle avoidance mode is an implicit one, i. e. the mode is not invoked by the user on purpose. Thus, the driver probably does not adapt to the new situation after an obstacle has been detoured, because he or she did not notice that the operation mode changed from operator-control to obstacle avoid-ance. It is very likely that the user would not react immediately after the avoidance maneuver and steer back to the original path. Instead, he or she would probably not change the joystick command. As a consequence, the wheelchair would follow the wrong track. This mode confusion problem motivates an additional feature of the obstacle avoidance algorithm: It is able to steer back to the heading of the original path after the obstacle has been passed. If the user does not adapt to the new situation, i. e. he or she does not change the joystick position after a detouring maneuver, the algorithm will interpret the command in the frame of reference that was current when the maneuver began. As a consequence, it is able to navigate through a corridor full of people or static obstacles by simply pointing forward with the joystick. If there is an object that has to be detoured, the user keeps the joystick in an unchanged position and thereby enables the obstacle avoidance al-gorithm to steer back to the orientation of the original path.

Please note that the narrow safety notion introduced here is extended during the following case study: Any behavior of the wheelchair that contributes to its obstacle avoidance skill is considered to be safety-relevant.

7

A theory has only the alternative of being right or wrong.
A model has a third possibility: it may be right, but irrelevant.
Manfred Eigen, 1973

Modeling Approach

7.1 Precise Modeling

Before mode confusion problems can be discussed, some remarks on modeling a technical system in general are necessary. The user of a running technical system has a strict *black-box view*. Since the user's problems are to be solved, the designer must take the same black-box point of view. This statement appears to be obvious, but has far-reaching consequences for the notion of mode. The user has no way of observing the current internal state, or mode, of the technical system.

Nevertheless, it is possible to describe a technical system entirely in a black-box view. The software engineering approach used here is inspired by the work of Parnas and Madey (1995) and Schouwen et al. (1993), even though it is based on events instead of Parna's variables.

An observer can observe (only) the environment of the technical system. When something relevant happens, this is called an *event*. When the technical system is the control unit of an automated wheelchair, then an event may be that the user pushes the joystick forward, that the wheelchair starts to move, or that the distance between the wheelchair and a wall ahead becomes smaller than the threshold of 70 cm.

The technical system has been constructed according to some *requirements* document REQ. It contains the requirements on the technical system, which is called SYSREQ, and those on the system's environment NAT. However, if an existing system is dealt with for which a requirements specification is no (more) available, it might be necessary to "reversely engineer" it from the implementation.

For the wheelchair, SYSREQ should state that the event of the wheelchair starting to move follows the event that the joystick is pushed forward. SYSREQ should also state what happens after the event of approaching a wall. Naturally, the wheelchair should not crash into a wall in front of it, even if the joystick is pushed forward. This can be described entirely in terms of observable events, by referring to the *history* of events until the current point in time. If the wheelchair has approached a wall, and if it has not yet moved back, it must not move further forward. For this description, no reference to an internal state is necessary.

Usually, several histories of events are equivalent with respect to what should happen in the future. For example, it does not matter on which path a wheelchair has approached a wall, as long as it stops to move. Such equivalences can greatly simplify the description of the behavior required, since only a few things about the history might need to be stated in order to characterize the situation.

In order to implement the requirements SYSREQ on a technical system, one usually needs several assumptions about the environment of the technical system to construct. For example, physical laws guarantee that a wheelchair will not crash into a wall ahead unless it has approached it closer than 70 cm and has continued to move for a certain amount of time. The documentation of assumptions about the environment is called NAT. NAT is part of the requirements document, too, in addition to the requirements on the technical system SYSREQ. NAT must be true even before the technical system is constructed. It is the implementer's task to ensure that SYSREQ is true provided that NAT holds.

7.2   Clarification of World Views

7.2.1   Where are the Boundaries?

The control software of a technical system cannot observe physical events directly. Instead, the technical system is designed such that sensor devices generate internal input events for the software, and the software's output events are translated by actuator devices back into physical events. Neither sensors nor actuators are perfectly precise and fast, therefore a distinct set of software events is used. Accordingly, the requirements on the technical system and the requirements on the software cannot be the same. For example, the wheelchair's ultrasonic distance sensors for the different directions can be activated in turns only, resulting in a noticeable delay for detecting obstacles.

The software requirements are called SOF, the requirements on the input sensors IN and the requirements on the output actuators OUT. Figure 7.1 on the next page shows the relationships among them. An important consequence is that the software SOF must compensate for any imperfectness of the sensors and actuators so that the requirements SYSREQ are satisfied. When defining SOF, it is necessary to specify whether the boundary of SOF or of SYSREQ is referred to.

This becomes even more important when the user who cooperates with the

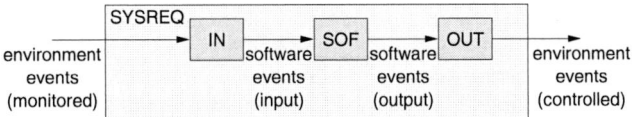

Figure 7.1: *System Requirements* SYSREQ *vs. Software Requirements* SOF.

technical system is considered. He or she observes the same events of the environment as the technical system does. But the user observes them through his/her own set of senses SENSE. SENSE has its own imperfections. For example, a human cannot see behind his/her back. The automated wheelchair will perceive a wall behind it when moving backwards, but the user will probably not. Therefore, what actually happens in reality (specified in REQ, i. e. the composition of SYSREQ and NAT) has to be distinguished from the user's mental model MMOD of it. When making a statement about MMOD, it is necessary to specify whether the boundary of MMOD or of REQ is referred to.

When defining the interfaces precisely, it turns out that there is an obvious potential that the software's perception of reality and the user's perception of it get out of synch. When analyzing this phenomenon, it is important to distinguish between three different interfaces:

- the *environment* to the machine (or to the user),

- the *software* to the input/output devices, and

- *mental* to the senses.

As a result, a precise relationship between reality as it is perceived by the user and his/her mental model of it can be established. This relation will be the basis of the definition of mode confusion.

The "direction" of an event is also distinguished, cf. 7.1. The software events of SOF are either *input events* or *output events*. The environment events of REQ are either *monitored events* or *controlled events*. The mental model MMOD uses a different set of *mental monitored events* and *mental controlled events*.

7.2.2   Brief Introduction to Refinement

As will be explained later, a particular kind of specification/implementation relation is used in the following sections. Such relations can be modeled rigorously by the concept of *refinement*. There exist a number of formalisms to express refinement relations. Here, CSP (Hoare, 1985) is used as specification language with the refinement semantics proposed by Roscoe (1997). One reason for this is that there is good tool support for performing automated refinement checks of CSP specifications thanks to the tool FDR (Roscoe, 1997). This section shall clarify the terminology for readers who are not familiar with the concepts.

In CSP, the behavior of a process P is described by the set $traces(P)$ of the event sequences it can perform. Since attention has to be paid to what can be done as well as to what can *not* be done, the traces model is not sufficient in this domain. It has to be enhanced by so-called *failures*.

Definition 7.1 (Failure) *A* failure *of a process P is a pair* (s, X) *of a trace s* ($s \in traces(P)$) *and a so-called* refusal *set X of events that may be blocked by P after the execution of s.*

If an output event o is in the refusal set X of P, and if there also exists a continuation trace s' which performs o, then process P may decide internally and non-deterministically whether o will be performed or not.

Definition 7.2 (Failure Refinement) *P* refines *S* in the failures model, written *$S \sqsubseteq_F P$, iff* $traces(P) \subseteq traces(S)$ *and also* $failures(P) \subseteq failures(S)$.

This means that P can neither accept an event nor refuse one unless S does; S can do at least every trace which P can do, and additionally P will refuse nothing more than S would. Failure refinement allows distinction between external and internal choice in processes, i.e. whether there is non-determinism. As this aspect is relevant for the application area, failure refinement is used as the appropriate kind of refinement relation.

7.2.3   Relation between Reality and the Mental Model

This approach is based on the motto *"The user must not be surprised"* as an important design goal for shared-control systems. Analogously, in the context of user interface design, Thimbleby (1990) refers to the "Principle of Least Astonishment" which states that a system, or its features, should be designed such that the user is surprised as little as possible. As a consequence, it has to be ensured that the perceived reality does not exhibit any behavior which cannot occur according to the mental model. Additionally, the user must not be surprised because something expected does *not* happen. When the mental model prescribes some behavior as necessary, reality must not refuse to perform it. These two aspects are described by the notion of failure refinement, as defined in the previous section.

There cannot be any direct refinement relation between a description of reality and the mental model, since they are defined over different sets of events (i.e., environment/mental). The user's senses SENSE should be understood as a relation from processes over environment events to processes over mental events. SENSE(REQ) is the user's perception of what happens in reality. The user is not surprised if SENSE(REQ) is a failure refinement of MMOD. As a consequence, the user's perception of reality must be in an *implementation/specification relationship* to the mental model.

Please note that an equality relation always implies a failure refinement relation, while the converse is not the case. If the user does not know how the system will behave with regard to some aspect, but knows that he/she does not know, then he/she will nevertheless experience no surprise. Such indifference can be expressed mathematically by a non-deterministic internal choice in the mental model.

7.2.4   Abstractions

When the user concentrates on safety, he/she performs an on-the-fly simplification of his/her mental model MMOD towards the safety-relevant part $MMOD_{SAFE}$. This helps him/her to analyze the current problem with the limited mental capacity. As Cañas et al. (2001) report, psychological studies show that users always adapt their current mental model of the technical system according to the specific task they carry out. The "initialization" of this adaptation process is the static part of their mental model, the so-called "conceptual model" (Sasse, 1997). This model represents the user's knowledge about the system and is stored in long term memory.

Analogously to the abstraction performed by the user, a simplification of the requirements document REQ to the safety-relevant part of it REQ_{SAFE} has to be performed. REQ_{SAFE} can be either an explicit, separate chapter of REQ, or it can be expressed implicitly by specifying an abstraction function, i. e. by describing which aspects of REQ are safety-relevant. REQ is abstracted out of three reasons: $MMOD_{SAFE}$ is defined over a set of abstracted mental events, and it can be compared to another description only if it is defined over the same abstracted set; the correctness of the safety-relevant part should be established without having to investigate the correctness of the entire mental model MMOD; and the model-checking tool support demands that the descriptions are restricted to certain complexity limits.

The abstraction functions are expressed mathematically in CSP by functions over processes. Mostly, such an abstraction function maps an entire set of events onto a single abstracted event. For example, it is irrelevant whether the wheelchair's speed is 81.5 or 82 cm/s when approaching an obstacle – all such events with a speed parameter greater than 80 cm/s will be abstracted to a single event with the speed parameter fast. Other transformations are hiding (or concealment (Hoare, 1985)) and renaming. But the formalism also allows for arbitrary trans-

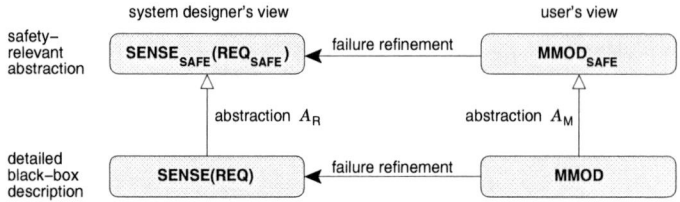

Figure 7.2: *Relationships Between the Different Refinement Relations.*

formations of behaviors; a simple example being a certain event sequence pattern mapped onto a new abstract event. The abstraction functions $\mathcal{A}_R$ and $\mathcal{A}_M$ are used for REQ and for MMOD, respectively.

The relation SENSE from processes over environment events to processes over mental events must be abstracted in an analogous way. It should have become clear by now that SENSE needs to be true, i. e. a bijection which does no more than some renaming of events. If SENSE is "lossy", we are already bound to experience mode confusion problems.

Figure 7.2 shows the relationships among the different descriptions. In order not to surprise the user with respect to safety, there must be a failure refinement relation on the abstract level between $\text{SENSE}_{\text{SAFE}}(\text{REQ}_{\text{SAFE}})$ and $\text{MMOD}_{\text{SAFE}}$, too.

8

A Rigorous View of Mode and Mode Confusion

This section presents new rigorous definitions of *mode* and *mode confusion*, and of *minimal safe mental model*. These definitions are then motivated and discussed.

8.1 Rigorous Definitions

In the following, let REQ_{SAFE} be a safety-relevant black-box requirements specification, let $SENSE_{SAFE}$ be a relation from processes over environment events to processes over mental events representing the user's senses, let $MMOD_{SAFE}$ be a safety-relevant mental model of the behavior of REQ_{SAFE}, and let Ev_{ctrl}^m be the set of mental controlled events which are not also mental monitored events.

Definition 8.1 (Potential future behavior) *A* potential future behavior *is a set of failures.*

Definition 8.2 (Mode) *A mode of* $SENSE_{SAFE}(REQ_{SAFE})$ *is a potential future behavior. And, a* mode *of* $MMOD_{SAFE}$ *is a potential future behavior.*

Definition 8.3 (Mode confusion) *A* mode confusion *between* $SENSE_{SAFE}(REQ_{SAFE})$ *and* $MMOD_{SAFE}$ *occurs if and only if* $SENSE_{SAFE}(REQ_{SAFE})$ *is not a failure refinement of* $MMOD_{SAFE}$ *(with hidden* Ev_{ctrl}^m*), i.e., iff*
$$MMOD_{SAFE} \setminus Ev_{ctrl}^m \not\sqsubseteq_F SENSE_{SAFE}(REQ_{SAFE}) \setminus Ev_{ctrl}^m .$$

Definition 8.4 (Minimal safe mental model) *The* minimal safe mental model *of the technical system T is the "smallest"* $MMOD_{SAFE}$ *for which the relation*

149

$\text{MMOD}_{\text{SAFE}} \setminus \text{Ev}^m_{\text{ctrl}} \quad \sqsubseteq_F \quad \text{SENSE}_{\text{SAFE}}(\text{REQ}_{\text{SAFE}}) \setminus \text{Ev}^m_{\text{ctrl}}$
and additionally the relation
$\text{SENSE}_{\text{SAFE}}(\text{REQ}_{\text{SAFE}}) \setminus \text{Ev}^m_{\text{ctrl}} \quad \sqsubseteq_F \quad \text{MMOD}_{\text{SAFE}} \setminus \text{Ev}^m_{\text{ctrl}} \text{ hold.}$

8.2   Reflection of Definitions

After the technical system T has moved through a history of events, it is in some "state". Since a black-box view has to be taken, two "states" can only be distinguished if T may behave differently in the future. The *potential future behavior* is described by a set of failures, such that both what T can do and what T can refuse to do are stated. This definition of "state" is rather different from the intuition in a white-box view, but necessarily so.

The next step to the notion of "mode" is more conventional. The notion of "state" is used here, if at all, in the context of the non-abstracted descriptions. Two states of a wheelchair are different, for example, if the steerable wheels will be commanded to a steering angle of 30 or 35 degrees respectively, within the next second. These states are equivalent with regard to the fact of obstacle avoidance. Therefore, both states are mapped to the same abstracted behavior by the safety-relevance abstraction function. Such a distinct safety-relevant potential future behavior is called a *mode*. Usually, many states of the non-abstracted description are mapped together to such a mode. On a formal level, both a state and a mode are a potential future behavior. The difference between both is that there is some important safety-relevant distinction between any two modes, which need not be the case for two states.

Parnas also defines a notion of mode (Schouwen et al., 1993). He uses it to simplify expressions in requirements documents. His notion of mode denotes an equivalence class of environmental states. This notion is therefore a specialization of his notion, since here such equivalence classes are only interesting with respect to safety-relevance. Parnas does not use safety-relevant abstractions as employed here. Additionally, Parnas partitions the state space completely into a set of modes, called a mode class. Also, he allows the use of more than one mode class at the same time. This is not necessary here.

After having clarified the notion "mode", the next step is to deal with *mode confusion*. The perceived reality and the user's mental model of it are in different modes at a certain point of time if and only if the perceived reality and the mental model might behave differently in the future, with respect to some safety-relevant aspect. Only if no such situation can arise in any possible execution trace, can it be said that there is no mode confusion. This means that the user's safety-relevant mental model must be a specification of the perceived reality. Expressed the other way around, the perceived reality must be an implementation of the user's safety-relevant mental model. This specification/implementation relationship can be described rigorously by failure refinement. If precise descriptions of both safety-relevant behaviors are available, it can be rigorously checked whether a mode confusion occurs. Since model-checking tool support exists, this check can even be automated. See Ch. 10 below.

Please note that the reality, as described by REQ, not only includes the system's requirements SYSREQ but also the environment requirements NAT. This restricts the behavior of SYSREQ by NAT: behavior forbidden by physical laws is not relevant for mode confusion.

When checking for failure refinement, any controlled event must be disregarded which is not also a monitored event. This is because the user cannot perceive such an event, and therefore cannot be surprised when an actual controlled event differs from the expected mental controlled event. As an example, consider the controlled event for a change of the target speed of the wheelchair. Such an event cannot be perceived by the user. Only a change in the actual speed (monitored event) will be perceived. Since this usually must happen sooner or later, this aspect has little practical relevance in most cases. The failure refinement check will fail anyway.

CSP provides the so-called "hiding operator ($\setminus$)" to specify the concept of disregarding a certain set of events for a process: A process $P \setminus E_h$ behaves as P does, except that each occurrence of an event from E_h in a trace of P is not observable for an external observer (i. e. it "is hidden"). In CSP, hiding is a standard means to abstract from the internal behavior of a process.

When concerned with mental models of the perceived reality, it is interesting to know the minimal knowledge which the user needs to have. Since a satisfied specification/implementation relationship between knowledge and the perceived reality is equivalent to the absence of a mode confusion problem, the first part of the definition of *minimal safe mental model* naturally builds on the definition of mode confusion, or of its absence, respectively. The first part alone guarantees minimality, but not safety. A model which "does not care at all" would satisfy this, i.e., one which allows and refuses all events non-deterministically. It is called *CHAOS* in CSP. The second part restricts the potential chaos and guarantees safe operation. The definition assumes a sensible relation $SENSE_{SAFE}$, which means that the user will notice any violation of a safety property at least eventually. The definition relies on the notion of a "size" of a behavior description. A precise definition of size is required for this. In this work, no such metrics are defined, but the above definition is rather provided as a template for anybody who discusses the size of models. In any case, the mutual refinement relation implies an observational equivalence of the minimal mental model with the safety-relevant requirements on the perceived reality, such that there are usually not too many candidates for comparison.

The mathematical description provided here allows for some interesting analysis of consequences. It is known in the literature that implicit mode changes may be a cause of mode confusion. In the description given here, an implicit mode change appears as an "internal choice" of the system, also known as a (spontaneous) "τ transition". The refinement relation dictates that any such internal choice must appear in the specification, too, which is the user's mental model in this case. This is possible: if the user expects that the system chooses internally between different behaviors, he/she will not be surprised, at least in principle. The problem is that the user must keep in mind all potential behaviors resulting from such a choice. If there is no clarifying event for a long time, the space of potential behaviors may grow very large and become impractical to handle in practice.

9

Beware of bugs in the above code;
I have only proved it correct, not tried it.
Donald Knuth, 1977

Methodological Results

The definitions given in Sect. 8 allow the classification of mode confusion problems, the derivation of recommendations from the classification for avoiding some of the problems, and the definitions can also be used as a foundation for detecting mode confusion.

9.1 Classification of Mode Confusion Problems

The clarification of world views in Sect. 7.2 makes it possible to *classify* mode confusion problems into three classes:

1. *Mode confusion problems which arise from an* incorrect observation *of the technical system or its environment.*

   Formally, this is the case when SENSE(REQ) is not a failure refinement of MMOD, but where SENSE(REQ) would be a failure refinement of MMOD, provided the user's senses SENSE would be a perfect mapping from environment events to mental events.

   The imperfections of SENSE may have *physical* or *psychological* reasons: either the sense organs are not perfect; for example eyes which cannot see behind the back. Or an event is sensed, but is not recognized consciously; for example because the user is distracted, or because the user currently is flooded with too many events. ("Heard, but not listened to.")

   Please note that the notion of mode confusion problem presented here also comprises the "knowledge gap" discussed in the research critique by Hourizi and Johnson (2001a,b) (see Sect. 6). Here, it appears as a mode confusion problem arising from an incorrect observation due to psychological reasons.

153

2. *Mode confusion problems which arise from the human's* incorrect knowledge *about the technical system or its environment.*

Formally, this is the case when SENSE(REQ) is not a failure refinement of MMOD, and when a perfect SENSE would make no difference.

3. *Mode confusion problems which arise from the* incorrect abstraction *of the user's knowledge to the safety-relevant aspects of it.*

Formally, this means that SENSE(REQ) is a failure refinement of MMOD, but $SENSE_{SAFE}(REQ_{SAFE})$ is not a failure refinement of $MMOD_{SAFE}$.

Since the safety-relevant requirements abstraction function $\mathcal{A}_R$ is correct by definition, the user's mental safety-relevance abstraction function $\mathcal{A}_M$ must be wrong in this case (compare Figure 7.2 above).

In contrast to previous classifications of mode confusion problems, this classification is *by cause* and not phenomenological, as, e.g., the one by Leveson et al. (1997). For example, mode confusion problems related to implicit mode changes are due to an incorrect abstraction, as long as the user knows about the possibility of the implicit mode change. If he/she does not know, the mode confusion is due to incorrect knowledge.

9.2 Recommendations for Avoiding Mode Confusion

The above causes of mode confusion problems lead directly to some *recommendations for avoiding* such problems. In order to avoid an incorrect observation of the technical system and its environment, it must be checked whether the user can physically observe all safety-relevant environment events. Furthermore, it must be checked whether the user's senses are sufficiently precise to ensure an accurate translation of these environment events to mental events. If this is not the case, then the system requirements have to be changed. An environment event controlled by the machine and observed by the user must be added. Such an event indicates the corresponding software input event. This measure has been recommended by others too, of course, but the rigorous view presented here now indicates more clearly when it must be applied.

Avoiding an incorrect observation also comprises that it is checked whether psychology ensures that observed safety-relevant environment events become conscious. The presented approach clearly points out the necessity of this check.

The check itself and any measures belong to the field of psychology, in which the author is not expert.

Establishing a correct knowledge of the user about the technical system and its environment can be achieved by documenting the requirements of them rigorously. This makes it possible to conceive user training material, such as a manual, which is complete with respect to functionality. This training material must not only be complete but also learnable. Complexity is an important learning obstacle. Therefore, the requirements of the technical system should allow as little non-deterministic internal choices as possible, since tracking all alternative outcomes is complex. This generalizes and justifies the recommendation by others to eliminate "implicit mode changes" (Degani et al., 1999; Leveson et al., 1997). Internal non-determinism may arise not only from the software, but also from the machine's sensor devices. If they are imprecise, the user cannot predict the software input events. Both kinds of non-deterministic internal choice can be eliminated by the same measure used against an incorrect physical observation: an environment event controlled by the machine is added which indicates the software's choice or the input device's choice, respectively.

Ensuring a correct mental abstraction process is mainly a psychological question and mostly beyond the scope of this paper. This work leads to the basic recommendation to either write an explicit, rigorous safety-relevant requirements document or to indicate the safety-relevant aspects clearly and separately in the general requirements document. The latter is equivalent to making the safety-relevance abstraction function for the machine $\mathcal{A}_R$ explicit. Either measure facilitates to conceive training material which helps the user to concentrate on safety-relevant aspects.

9.3 Automated Mode Confusion Detection

The definitions given in Sect. 8.1 form a *foundation for detecting* mode confusion by model checking. This foundation has opened new possibilities for a comprehensive analysis of mode confusion problems. The next chapter reports on some practical experience with this.

156

10

Practical Results: Wheelchair Case Study

This section illustrates the application of the approach to a real world scenario: the shared-control obstacle avoidance module of the Bremen Autonomous Wheelchair, as introduced in Sect. 6.2.

10.1 Methodology

The motion behavior of the wheelchair robot was specified in several steps. Both the requirements on the technical system and a mental model of it were specified. For both these requirements, a safety-relevant abstraction was also produced.

An explicit mental model in CSP was obtained through one of the three alternatives discussed in the introduction: a user interview. The author played the role of the "interviewed" user. He has built a mental model of the wheelchair robot through extensive use. Even though he knows the source code of the software, he definitely does not use this knowledge while driving. Instead, he has built his own, intuitive, black-box mental model. This model is much easier to use for quick decisions. It also turned out that this model is structured differently than the technical system. In addition, the user interviewed can express himself in CSP directly.

The CSP requirements specification of the wheelchair robot was extracted through "reverse engineering" from the source code. Unfortunately, no requirements document existed before this. Of course, this was done only *after* the mental model was specified, in order not to spoil the latter by a fresh and close impression of the code. The CSP specification is close to the source code. It was restricted to those parts related to motion. The rather complex sensor software is included at a high level of abstraction only.

Following the approach outlined in Sect. 7.1 and in Sect. 7.2.1 the mental model and the requirements were specified. Firstly, the boundary between the technical system and its environment was identified, along with the variables relevant at this boundary. These are the position of the joystick, the actual status of the wheelchair motors, the orientation of the wheelchair in the initial inertial system, the locations of obstacles, and the current command to the wheelchair motor systems. The latter is the only controlled variable, the others are monitored variables. Next, events were specified in CSP. These events denote a change in one of the variables. Finally, the mental model and the requirements were specified in CSP. At a few points, the process had to be iterated. The selection of relevant variables depends on the needs of the relations.

The boundary between the software and the input/output devices had to be identified for the specification of the software requirements SOF. The input and output devices comprise software too, the driver software. Therefore, there was some choice in drawing the boundary. Finally, the boundary was drawn such that the description became as simple as possible. In particular, the driver software of the sonar system provides "virtual sensors" (Lankenau and Röfer, 2001a) which allow the other software to inspect a virtual map of the obstacle situation. The input relation IN therefore became a nearly trivial mapping of obstacle locations to virtual obstacle locations with this respect.

It was surprisingly little work to produce the safety-relevant abstractions of the mental model and of the requirements specification afterwards. First, the detailed data types that describe measured values and motor commands were considered. For them, it was documented explicitly, which properties are safety-relevant. Then, the detailed data types were replaced by suitably abstracted ones. The latter only have two to three or sometimes four distinct values instead of some integer ranges. For example, an integer-valued speed command range between −42 cm/s and 84 cm/s was abstracted to the three values standSt, slowSpeed, and fastSpeed. They cover all distinct safety-relevant cases. It is not even distinguished between forward and backward driving, since the setting turned out to be symmetrical with this respect.

The most difficult abstraction was that of the virtual map of the obstacle situation. Only the closest obstacle on the current path of the wheelchair is kept. An object in the surrounding is a relevant obstacle if driving further on the current path would cause some part of the wheelchair to collide with the object. The position of the obstacle relative to the wheelchair is described by a potential wheelchair path and a distance. The path is defined by the steering angle that would be necessary for a collision of the center of the wheelchair's front axle with the obstacle. The distance is the travel distance before impact. Since the distance as such is not interesting here, it is abstracted to the corresponding criticality with respect to the current wheelchair speed: if the obstacle is far away, the required action is less

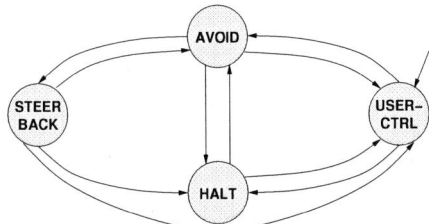

Figure 10.1: *The User's Mental Model of the Wheelchair Represented by four CSP Processes.*

demanding than it is if the obstacle is close.

The behavior description part of the CSP specifications needed very few modifications. All motion-related behavior is potentially safety-relevant. But with other applications, more simplifications might be possible and necessary.

10.2 Overview of CSP Specifications

There are four specifications in CSP as shown in Fig. 7.2 on page 148. They partially share definitions of events and types as appropriate. Ultimately, they are combined for a refinement check of the detailed descriptions on the one hand, and for an automated refinement check of the safety-relevant abstractions on the other . As expected, an automated refinement check at the detailed level is not possible since the state space is way too large.

The user's mental model of the wheelchair's obstacle avoidance module (see Sect. 6.2) is specified by four major CSP processes: a "halt" process entered whenever the joystick is in the neutral position, a "user controlled" process for user controlled driving, an "avoid" process for the obstacle avoidance skill of the wheelchair, and a "steer-back" process in which the wheelchair automatically returns to the original driving direction after an obstacle avoidance maneuver has been completed. Figure 10.1 shows the processes and the transitions among them.

The structure of the specification of the wheelchair requirements and of the mental model of it are different. The specification of the wheelchair requirements REQ is the composition of the requirements on the technical system SYSREQ and of the requirements on its environment NAT. SYSREQ in turn is a composition of IN, SOF, and OUT as shown in Fig. 7.1 on page 145. SOF performs an infinite loop of reading sensors, choosing the appropriate software routine, processing the input, and setting the actuators. The specification of the mental model MMOD is the composition of the mental model of the technical system SYSMMOD and of the mental model of its environment NATMMOD. SYSMMOD performs a loop of perception, calculation, and acting. There are no separate input/output relations, and the detailed structure of SYSMMOD is also quite different from that of SOF. For each of the above, the structure of the detailed and of the abstracted specifications are identical.

The size of both safety-relevant abstractions together is about 1200 lines of commented CSP specification, or about 620 lines of pure CSP specification. The size of the non-abstracted specifications is a little less, since they were specified with less details of the auxiliary functions in the expectation that model checking would be impossible here anyway.

10.3   Mode Confusions Detected During Modeling

The rigorous modeling process revealed a mode confusion problem which arises from an incorrect observation of the environment by the user. The user's senses SENSE had to be specified explicitly. SENSE does a direct one-to-one mapping of monitored events to mental monitored events (with some delay). But the explicit modeling made it obvious that there is one exception. The user cannot see obstacles behind his back. Of course, this is already a problem when driving backward. But they are likely to obstruct forward paths, too: when driving a curve, the back of the wheelchair swerves out to the side and may hit obstacles which are nearby alongside the wheelchair, but behind the user's head. The wheelchair robot will notice the danger, activate the obstacle avoidance skill, and change the motion into a safe one. The user will not notice the mode change, and he or she will be surprised. This is the reason why driving backward has the exactly same problems as driving forward, and can be ignored out of symmetry considerations in the abstraction, as mentioned above.

This mode confusion problem can be resolved by adding a feedback light to the user interface. It is on when the system is in the "avoid" or "steer-back" software routine.

10.4   Mode Confusions Detected by Model Checking

An automated analysis detected three mode confusion problems which were new. This happened despite the author knowing the wheelchair robot well, even its more obscure properties. Additionally, the automated analysis detected all expected mode confusion problems.

The first new mode confusion problem occurs when the different senses of the user work at different speeds. The relation $SENSE_{SAFE}$ translates monitored events to mental monitored events. In a first version, this translation was specified independently for each of the user's senses (vision, tactile, motion-detection). This is realistic, since the inner ear can take some time before detecting a slow turning, and since the user might not see an obstacle in a complex surrounding immediately. These delays need not be correlated. And the joystick position is felt practically without delay by the user.

The automated model-checking tool FDR (Roscoe, 1997) detected a violation of the refinement property in Def. 8.3 resulting from this. It generated an example trace of events up to a position where the perceived technical system and the mental model behaved differently. An inspection of this example showed the cause. The problem was resolved in the case study by adding the explicit assumption that the user's senses will not delay events noticeably.

The mode confusion problem found will occur in most shared-control systems. It occurs if the user uses different senses, and if these can have different delays for perception. With this respect, a complex visual scene can already count as being perceived by different senses, like the collection of aircraft cockpit panels.

In the classification given in Sect. 9.1, the problem found is one arising from an incorrect observation of the environment by the user.

The second mode confusion problem revealed by the FDR tool is caused by an erroneous simplification of the acquired mental model by the user. Rushby (2001) denotes this process of irregularly generalizing often used knowledge as *inferential simplification*. The author simplified his model of the "halt" routine (see above) such that the wheelchair was assumed to re-set the steering angle to its initial "straight" position whenever the intended user speed was set to zero. As a consequence, according to this mental model the wheelchair could not change its steering to a value other than `straight` when standing still. But the technical system allows this.

The FDR tool reported the refinement violation shown in Tab. 10.1: the user

Table 10.1: *Traces Leading to the Detected Mode Confusion.* Mental model as "specification" (left column) vs. perceived reality as "implementation" (right column).

$\text{MMOD}_{\text{SAFE}} \setminus \text{Ev}^{m}_{\text{ctrl}}$		$\not\sqsubseteq_F$	$\text{SENSE}_{\text{SAFE}}(\text{REQ}_{\text{SAFE}}) \setminus \text{Ev}^{m}_{\text{ctrl}}$:
mmJoystickCmd.0.left	$\downarrow$		mmJoystickCmd.0.left
			mmMotorsActual.standSt.left

intends to steer to the left while the wheelchair stands still. Both, the perceived reality as well as the mental model of the reality engage in the corresponding mental monitored event mmJoystickCmd.0.left. The technical system correctly maps this joystick command to the corresponding motor command (mental controlled event that is not visible in the trace in Tab. 10.1). As a consequence a change in the actual motors status is perceived as the mental monitored event `mmMotorsActual.standSt.left`. Due to the inferential simplification mentioned above, the mental model refuses to engage in this event, it only accepts `mmMotorsActual.standSt.straight` here.

Please note that this mode confusion is safety-relevant: if the user's mental model does not allow to change the steering angle during a standstill, the user might lose track of the automation behavior: consider a situation in which it is necessary to set the steering angle to its maximum value on either side to avoid a certain object in front of the wheelchair. If your mental model refuses to steer while standing still, you might not be able to set the steering angle soon enough while driving. This is because the curve radius increases when you steer while you are already driving (even at a very low speed level). Therefore, this mode confusion may decide about whether or not it is possible to pass an obstacle, and is thus safety-relevant.

In the above classification, this mode confusion results from incorrect knowledge of the user about the system caused by an erroneous inferential simplification.

This mode confusion was resolved by refreshing the author's knowledge about the "halt" routine: The corrected version of his mental model allows to change the steering angle while the wheelchair is in a standstill. This enhanced version of the mental model is used in the following.

By chance, the third mode confusion detected by the FDR tool falls into the third category of the classification given in Sect. 9.1: it is caused by a wrong safety-relevant abstraction of the user's mental model. Please note that even though the three mode confusions detected with the help of the FDR support perfectly cover the three categories of causes for mode confusions, this is *not* constructed but simply what was observed.

This mode confusion occurs due to an erroneous abstraction of specific information required for the decision whether the obstacle avoidance maneuver should be pre-empted by the user. As mentioned in Sect. 6.2, the user may regain the control of the wheelchair during the automatic obstacle avoidance maneuver by moving the joystick. The detailed version of the mental model "stores" the previous setting of the joystick and always compares it to a new setting. The safety-relevant abstraction follows the correct idea that the exact previous command is not safety-relevant. The important information is whether or not the user moved the joystick. The handling of such a change in the joystick position was erroneously represented in the abstracted version of the author's mental model. As a result, he might be confused during an avoidance maneuver, as shown in Tab. 10.2. After having passed the critical obstacle on the user-intended left side, the situation clarifies as there are no more relevant obstacles. This can be observed with the help of the mental monitored event `mmObsChange.straight.cAngDist`. As the

Table 10.2: *Traces Leading to the Detected Mode Confusion.* Mental model (left column) vs. reality perceived by "perfect" senses (right column).

$$\text{MMOD}_{\text{SAFE}} \setminus \text{Ev}^m_{\text{ctrl}} \quad \not\sqsubseteq_F \quad \text{SENSE}_{\text{SAFE}}(\text{REQ}_{\text{SAFE}}) \setminus \text{Ev}^m_{\text{ctrl}}:$$

MMOD$_{\text{SAFE}} \setminus$ Ev$^m_{\text{ctrl}}$		SENSE$_{\text{SAFE}}$(REQ$_{\text{SAFE}}) \setminus$ Ev$^m_{\text{ctrl}}$:
mmJoystickCmd.100.left		mmJoystickCmd.100.left
mmObsChange.straight.cAngDist	↓	mmObsChange.straight.cAngDist
mmObsChange.left.nonCritDist		mmObsChange.left.nonCritDist
	↓	mmMotorsActual.fastSpeed.right

joystick has not been moved since the avoidance maneuver began, the wheelchair steers back to the original orientation, i. e. it turns to the right. Due to the error in the abstraction, the mental model cannot follow this path. It erroneously assumes the joystick to have been moved and passes the joystick command to the motor. Thus, it refuses the mental monitored event mmMotorsActual.fastSpeed.right. Instead, it expects mmMotorsActual.fastSpeed.left.

The automated analysis also detected the mode confusion problem which was already detected during modeling. The relation SENSE$_{\text{SAFE}}$ was specified such that the visual perception of the closest obstacle may be replaced by the perception of a less critical one. The model-checking tool generated an example trace where the wheelchair appeared to change its motion behavior without a cause, as shown in Tab. 10.3. The user erroneously assumes an object located on a potential path

Table 10.3: *Traces Leading to the Detected Mode Confusion.* Mental model (left column) vs. reality perceived by "lossy" (wrong obstacle) senses (right column).

$$\text{MMOD}_{\text{SAFE}} \setminus \text{Ev}^m_{\text{ctrl}} \quad \not\sqsubseteq_F \quad \text{SENSE}^{obs}_{\text{SAFE}}(\text{REQ}_{\text{SAFE}}) \setminus \text{Ev}^m_{\text{ctrl}}:$$

MMOD$_{\text{SAFE}} \setminus$ Ev$^m_{\text{ctrl}}$		SENSE$^{obs}_{\text{SAFE}}$(REQ$_{\text{SAFE}}) \setminus$ Ev$^m_{\text{ctrl}}$:
mmObsChange.right.stopDist		mmObsChange.right.stopDist
mmJoystickCmd.100.straight	↓	mmJoystickCmd.100.straight
		mmMotorsActual.standSt.straight

to the right to be the closest obstacle in a certain situation. He is not aware that there is a closer object in front of the wheelchair. Therefore he is confused when the intended driving command issued via the joystick (reported by the mental monitored event mmJoystickCmd.100.straight) cannot be realized because of the obstacle straight in front of the wheelchair. The user's mental model refuses to engage in the event denoting that the wheelchair reduced its speed to a standstill because it assumes the closest obstacle to be on an irrelevant path to the right.

The resulting mode confusion is very common in manned robotics: the human driver of the robot (here: the wheelchair) is not correctly aware of the obstacle situation in the surrounding of the wheelchair. As a consequence, the user is surprised if the automation intervenes where there seems to be no reason for such an intervention. Or, vice versa, the user cannot track the automation's behavior if it does *not* intervene while the user expects it should do so. In the above classification, such a mode confusion arises from an incorrect observation of the environment: the "wrong" obstacle is assumed to be the closest obstacle to be detoured.

The next iteration of automated analysis based on the enhanced version of the mental model proved that there is *no further mode confusion*, except those arising from an incorrect observation of the environment by the user. An "ideal" version $\text{SENSE}_{\text{SAFE}}^{ideal}$ of $\text{SENSE}_{\text{SAFE}}$ with no delay and no modification of events was used for this analysis. The model-checking tool investigated all traces of events theoretically possible and thereby conducted a mathematical proof by exhaustive enumeration. The size of the state space to be explored during one of the refinement checks is in the order of 100,000 states.

As shown, this case study proves the feasibility of the approach. When a professional psychologist obtains an explicit mental model from a less expert user, this will have more mode confusion problems, and the automated analysis must find them. This is demonstrated with a made-up mental model. This mental model is the same as the previous one, but lacks the obstacle avoidance skill of the wheelchair. In this experiment, the user thinks he or she is always in control. The CSP specification of the corresponding "simplistic" mental model comprises only two processes, "user controlled" and "halt". The model checking tool generated an example error trace for this configuration where the wheelchair's motion is inconsistent with the mental model, see Tab. 10.4. After having engaged in the

Table 10.4: *Traces Leading to the Detected Mode Confusion.* Simplistic mental model (left column) vs. reality perceived by "ideal" (right column).

$\text{MMOD}_{\text{SAFE}}^{simple} \setminus \text{Ev}_{\text{ctrl}}^{\text{m}}$	$\not\sqsubseteq_F$	$\text{SENSE}_{\text{SAFE}}^{ideal}(\text{REQ}_{\text{SAFE}}) \setminus \text{Ev}_{\text{ctrl}}^{\text{m}}$:
mmJoystickCmd.100.left		mmJoystickCmd.100.left
mmObsChange.right.stopDist	$\downarrow$	mmObsChange.right.stopDist
		mmMotorsActual.standSt.left

mental monitored events to read the current joystick command and to perceive the obstacle situation, the simplistic mental model refuses to accept the event mmMotorsActual.standSt.left reporting that the wheelchair reduced its speed to a standstill. The reason for the braking maneuver is the detected obstacle that appeared very close to the wheelchair. The technical system reacts as required by setting the target speed to zero (corresponding controlled event not observable in the trace). The simplistic mental model fails to follow this speed reduction and still sets the target speed to the user intended value (again, the corresponding controlled event is not observable here). As a result, the expected mode confusion occurs.

In order to show some more examples of error traces, experiments were conducted with two other realistic situations dealing with an imperfect user perception.

In the first scenario, it is assumed that the user's senses are perfect, except for his ability to estimate the wheelchair's current speed. That means, the user can only distinguish between a standstill and a movement, but he may mix up slow and fast motion. The author's mental model that has already been proven to be free from mode confusions is used here. But the user is forced to perceive the reality through his imperfect senses. The mode confusion shown in Tab. 10.5 was found by the model-checking tool. Since the current speed of the wheelchair is

Table 10.5: *Traces Leading to the Detected Mode Confusion.* Mental model (left column) vs. reality perceived by "lossy" (imperfect perception of wheelchair speed) senses (right column).

$$\text{MMOD}_{\text{SAFE}} \setminus \text{Ev}_{\text{ctrl}}^m \quad \not\sqsubseteq_F \quad \text{SENSE}_{\text{SAFE}}^{speed}(\text{REQ}_{\text{SAFE}}) \setminus \text{Ev}_{\text{ctrl}}^m :$$

MMOD_SAFE \ Ev	SENSE_SAFE
mmJoystickCmd.100.left	mmJoystickCmd.100.left
↓	mmMotorsActual.slowSpeed.left

safety-relevant, a deviation of the perceived speed and the one expected by the mental model would cause a mode confusion. The FDR tool finds that after having engaged in the mental monitored event mmJoystickCmd.100.left, the mental model process refuses to engage in the mental monitored event prescribed by the perceived reality. This is because the *real* speed of the wheelchair is fastSpeed, but the imperfect senses perceive it as slowSpeed. Thus, the expected mode confusion is found.

The second scenario is a slight but meaningful alteration of the example mentioned above. This time, the user is assumed to employ the adequate author's mental model (which is mode confusion free, as shown above). The user has perfect senses with the exception that he or she may fail in estimating the correct distance to a detected obstacle. If a mode confusion is found here, it shows that it does not suffice to detect an obstacle, but it is as important to be able to estimate the travel distance between the wheelchair and this obstacle. Table 10.6 shows that the FDR tool finds such a mode confusion. The error traces directly show the problem: the user detects an obstacle in front of the wheelchair, on the left-hand side. Even though the distance to the obstacle allows to detour the obstacle by an appropriate avoidance maneuver, the user perceives the object as if it were very close to the wheelchair. In his or her opinion the only adequate reaction would be to stop immediately. As a consequence, the user is surprised (mode confusion) when the wheelchair does *not* stop but passes the obstacle on its left side.

Table 10.6: *Traces Leading to a Detected Mode Confusion.* Mental model (left column) vs. reality perceived by "lossy" (correct obstacle, but maybe wrong distance estimate) senses (right column).

$\mathrm{MMOD_{SAFE} \setminus Ev^m_{ctrl}}$	$\not\sqsubseteq_F$	$\mathrm{SENSE^{dist}_{SAFE}(REQ_{SAFE}) \setminus Ev^m_{ctrl}}$:
`mmJoystickCmd.100.left`		`mmJoystickCmd.100.left`
`mmObsChange.left.stopDist`	$\downarrow$	`mmObsChange.left.stopDist`
		`mmMotorsActual.slowSpeed.left`

10.5 Minimal Safe Mental Model

During the work on the case study, an interesting observation on the minimality of the mental model used was made. The mental model extracted by the "interview" of the author is represented by four major CSP processes. They correspond to the four software routines of the obstacle avoidance module. When using the "ideal" version of the user's senses, it was proved that the perceived behavior of the wheelchair is a failures refinement of the mental model of it, and vice versa. According to Def. 8.4 on page 149, this mental model could therefore be a minimal safe mental model, provided that there is no "smaller" one satisfying this criterion.

It turned out that a smaller safe mental model exists. While working on the specifications, the first author of Bredereke and Lankenau (2002) built his own mental model, which can be represented by only three CSP processes. The only difference to the author's mental model is that the original "halt" process has become part of an extended "user controlled" process. In other words, any transition to the original halt process does now lead to this new user controlled process. As a result, the specification has become simpler. An automated refinement check proved that it still satisfies the criterion from Def. 8.4. This means, that the reduced mental model is still equivalent to the perceived reality in the failures model. It is therefore still without mode confusion problems and also still safe.

This shows that the rigorous definition of a minimal safe mental model can have a practical value. The aspect that is still missing is a notion of "size". In this case, the size comparison result is obvious. For the general case, one will need an explicit metrics.

11

. . . , and I have assuredly found an admirable proof of this,
but the margin is too narrow to contain it.
P. de Fermat, 1621

Summary and Future Work

This thesis part presents a rigorous way of modeling the user and the machine in a shared-control system. This makes it possible to propose precise definitions of "mode" and "mode confusion" , and of "minimal safe mental model".

In the modeling approach, the commonly used distinction between the machine and the user's mental model of it is extended by explicitly separating these and their safety-relevant abstractions. Furthermore, it is shown that distinguishing three different interfaces during the design phase reduces the potential for mode confusion.

The proposition that the user must not be surprised leads directly to the conclusion that the relationship between the mental model and the machine must be one of specification to implementation, in the mathematical sense of refinement. Mode confusion can occur if and only if this relation is not satisfied.

A result of this insight is a new classification of mode confusion by cause, leading to a number of design recommendations for shared-control systems which help to avoid mode confusion problems.

Since tools to model-check refinement relations exist, this approach supports the automated detection of remaining mode confusion problems.

Those model-checking tools are most suitable for detecting mode confusion problems, which allow a description of the specification as an explicit model. This is because the mental model is an explicit model. The other kind of tools existing allow the description of the specification in a property-oriented way, making its model implicit. This explains why Buth (2002) found that CSP and FDR were most suitable for detecting mode confusion problems (see Sect. 6.1.2).

The practical feasibility of the approach is demonstrated in a case study on a wheelchair robot. The rigorous modeling process presented here already revealed a mode confusion problem. The automated analysis detected three other mode confusion problems, which were new. The automated analysis then proved mathematically that there is no further mode confusion possible in the case study. Finally, a smaller mental model was found and could be proven to be still a safe mental model.

The proposed process for tackling mode confusion can be summarized as follows: 1. describe rigorously, 2. avoid problems by following the recommendations above, and 3. detect the remaining problems by model checking.

This work lends itself to extension into several directions. The recommendations for avoiding mode confusion problems should be verified experimentally. Experts in psychology will be able to implement the non-technical rules given above by concrete measures. The definition template for a "minimal safe mental model" can be instantiated with suitable metrics for the size of a mental model. Finally, there are still more application domains beyond aviation and robotics.

12

Arthur looked up. "Ford!" he said, "there's an infinite number of monkeys
...who want to talk to us about this script for Hamlet they've worked out."
"The Hitchhiker's Guide to the Galaxy", Douglas Adams

Publications of the author

This thesis is based on and extends material that has already been published in workshop notes, conference proceedings, journals, and books (listed in chronological order):

1. **J. Kollmann, A. Lankenau, A. Bühlmeier, B. Krieg-Brückner, and T. Röfer.** Navigation of a kinematically restricted wheelchair by the parti-game algorithm. In *Spatial Reasoning in Mobile Robots and Animals*, pages 35 – 44, Manchester University, 1997. AISB-97 Workshop.

2. **A. Lankenau and O. Meyer.** Der autonome Rollstuhl als sicheres eingebettetes System. Master's thesis, Fachbereich Mathematik und Informatik, Universität Bremen, 1997.

3. **A. Lankenau, O. Meyer, and B. Krieg-Brückner.** Safety in robotics: The Bremen Autonomous Wheelchair. In *Proceedings of AMC'98, 5th Int. Workshop on Advanced Motion Control*, pages 524 – 529, Coimbra, Portugal, 1998.

4. **A. Lankenau and T. Röfer.** Architecture of the Bremen Autonomous Wheelchair. In B. Hildebrand, R. Moratz, and Ch. Scheering, editors, *Architectures in Cognitive Robotics. Technischer Bericht 98/13*, pages 19 – 24. SFB 360 – Situierte Künstliche Kommunikatoren. Universität Bielefeld, 1998.

5. **T. Röfer and A. Lankenau.** Architecture and applications of the Bremen Autonomous Wheelchair. In P. P. Wang, editor, *Proc. of the Fourth Joint Conference on Information Systems*, volume 1, pages 365 – 368. Association for Intelligent Machinery, 1998.

6. **A. Lankenau and O. Meyer.** Formal methods in robotics: Fault tree based verification. In *Proc. of Quality Week Europe*, Brussels, Belgium, 1999.

7. **T. Röfer and A. Lankenau.** Ensuring safe obstacle avoidance in a shared-control system. In J. M. Fuertes, editor, *Proc. of the 7th Int. Conf. on Emergent Technologies and Factory Automation*, pages 1405 – 1414, 1999.

8. **T. Röfer and A. Lankenau.** Ein Fahrassistent für ältere und behinderte Menschen. In *Autonome Mobile Systeme 1999*, Informatik aktuell, pages 334 – 343, Berlin, Heidelberg, New York, 1999. Springer.

9. **R. Müller, T. Röfer, A. Lankenau, A. Musto, K. Stein, and A. Eisenkolb.** Coarse qualitative descriptions in robot navigation. In C. Freksa, W. Brauer, C. Habel, and K.F. Wender (editors): *Spatial Cognition II*, Lecture Notes in Artificial Intelligence (Vol. 1849), pages 265–276, Berlin, Heidelberg, New York, 2000. Springer.

10. **T. Röfer and A. Lankenau.** Architecture and applications of the Bremen Autonomous Wheelchair. *Information Sciences*, Elsevier. 126(1-4):1 – 20, Jul. 2000.

11. **A. Lankenau and T. Röfer.** Smart Wheelchairs - State of the Art in an Emerging Market. *Künstliche Intelligenz (Heftschwerpunkt Autonome Mobile Systeme)*, (4):37 – 39, 2000. Gesellschaft für Informatik e.V., arenDTaP Verlag.

12. **A. Lankenau and T. Röfer.** Rollstuhl "Rolland" unterstützt ältere und behinderte Menschen. *FIfF-Kommunikation "Informationstechnik und Behinderung"*, (2):48 – 50, 2000. Forum InformatikerInnen für Frieden und gesellschaftliche Verantwortung (FIfF) e.V.

13. **A. Lankenau and T. Röfer.** The Role of Shared Control in Service Robots – The Bremen Autonomous Wheelchair as an Example. In: Workshop Notes "Service Robotics – Applications and Safety Issues in an Emerging Market", Berlin, Germany, 2000. European Conference on Artificial Intelligence (ECAI 2000). Pages 27 – 31.

14. **T. Röfer, A. Lankenau, and R. Moratz, editors.** *Workshop "Service Robotics – Applications and Safety Issues in an Emerging Market"*, Berlin, Germany, 2000. European Conference on Artificial Intelligence (ECAI 2000).

15. **A. Lankenau.** Avoiding mode confusion in service-robots. In M. Mokhtari, editor, *Integration of Assistive Technology in the Information Age, Proc. of the 7th Int. Conf. on Rehabilitation Robotics*, pages 162 – 167, Evry, France, Apr. 2001. IOS Press.

16. **A. Lankenau and T. Röfer.** The Bremen Autonomous Wheelchair – a versatile and safe mobility assistant. *IEEE Robotics and Automation Magazine, "Reinventing the Wheelchair"*, 7(1):29 – 37, Mar. 2001a.

17. **A. Lankenau and T. Röfer.** Selbstlokalisation in Routengraphen. In P. Levi and M. Schanz, editors, *Autonome Mobile Systeme 2001*, Informatik aktuell, pages 157 – 163. Springer, 2001b.

18. **A. Lankenau and T. Röfer.** Mobile robot self-localization in large-scale environments. In *Proc. of the Int. Conf. on Robotics and Automation ICRA 2002*, Washington, D.C., USA, 2002. IEEE, Omnipress. 1359 – 1364.

19. **A. Lankenau, T. Röfer, and B. Krieg-Brückner.** Self-localization in large-scale environments for the Bremen Autonomous Wheelchair. In C. Freksa, W. Brauer, C. Habel, and K.F. Wender (editors) *Spatial Cognition III*, Lecture Notes in Artificial Intelligence (Vol. 2685), Berlin, Heidelberg, New York, 2002. Springer.

20. **J. Bredereke and A. Lankenau.** A Rigorous View of Mode-Confusion. In: S. Anderson, S. Bologna, and M. Felici (eds.): *Computer Safety, Reliability and Security, Proc. of the 21st Int'l Conf. SafeComp 2002*, Catania, Italy. Lecture Notes in Computer Science, Vol. 2434 Berlin, Heidelberg, New York, 2002. Springer. 19 – 31.

21. **T. Röfer and A. Lankenau.** Route-Based Robot Navigation. In: Freksa, C. (Hrsg.): *Künstliche Intelligenz (Themenheft Spatial Cognition)*, 2002. Fachbereich 1 der Gesellschaft für Informatik e.V., arenDTaP Verlag. 29 – 31.

Bibliography

E. Abbott and D. Powell. Land Vehicle Navigation Using GPS. *Proceedings of the IEEE, Special Issue on Global Positioning System*, 87(1):145 – 162, 1999.

ACM. Robot wheelchair climbs stairs. *Communications of the ACM*, 43(2):10, 2000.

K.O. Arras and N. Tomatis. Improving Robustness and Precision in Mobile Robot Localization by Using Laser Range Finding and Monocular Vision. In *Proc. of the 3rd European Workshop on Advanced Mobile Robots (Eurobot '99)*, Zurich, Switzerland, September 1999.

T. Bachmann and S. Bujnoch. ConnectedDrive - Driver Assistance Systems of the Future. In *Commercial Applications of Satellite-Navigation*, 1999.

D. Bank. A High-Performance Ultrasonic Sensing System for Mobile Robots. In *Robotik 2002*, number 1679 in VDI-Berichte, pages 557 – 564, Ludwigsburg, Germany, jun 2002. VDI/VDE-Gesellschaft Mess- und Automatisierungstechnik, VDI-Verlag GmbH, Düsseldorf.

J. Baus, A. Krüger, and W. Wahlster. A resource adaptive mobile navigation system. In *Proc. of the Int.'l Conf. on Intelligent User Interfaces (IUI 2002)*, San Francisco, USA, January 2002. ACM, ACM Press, New York, USA.

H. Behrens. Info from TeleAtlas AG, Germany. phone call on June, 27th, 2002.

L.M. Bergasa, M. Mazo, A. Gardel, J.C. García, A. Ortuño, and A.E. Méndez. Guidance of a wheelchair for handicapped people by face tracking. In J. M. Fuertes, editor, *Proc. of the 7th Int. Conf. on Emergent Technologies and Factory Automation*, pages 105 – 111, Barcelona, Spain, 1999.

D. Bernstein and A. Kornhauser. An introduction to map matching for personal navigation assistants. Technical report, Princeton University, August 1996.

J. Borenstein, H. R. Everett, and L. Feng. *Navigating Mobile Robots – Systems and Techniques.* A.K. Peters, Ltd., USA, 1996.

J. Borenstein and Y. Koren. The vector field histogram - fast obstacle avoidance for mobile robots. *IEEE Journal on Robotics and Automation*, 7(3):278–288, 1991.

B. Borgerding, O. Ivlev, C. Martens, N. Ruchel, and A. Gräser. FRIEND — functional robot arm with user friendly interface for disabled people. In *Proc. of the 5th European Conf. for the Advancement of Assistive Technology*, Düsseldorf, Germany, 1999.

J. Bredereke and A. Lankenau. A rigorous view of mode confusion. In S. Anderson, S. Bologna, and M. Felici, editors, *Proc. of Safecomp 2002, 21st Int'l Conf. on Computer Safety, Reliability and Security*, number 2434 in LNCS, pages 19 – 31, Catania, Italy, September 2002. Springer.

W. Burgard, D. Fox, and D. Henning. Fast grid-based position tracking for mobile robots. In G. Brewka, Ch. Habel, and B. Nebel, editors, *KI-97: Advances in Artificial Intelligence*, Lecture Notes in Artificial Intelligence, pages 289–300, Berlin, Heidelberg, New York, 1997. Springer.

B. Buth. *Formal and Semi-Formal Methods for the Analysis of Industrial Control Systems*, volume 15 of *BISS Monographs*. 2002. (Habilitationsschrift submitted May 2001).

R.W. Butler, S.P. Miller, J.N. Pott, and V.A. Carreño. A formal methods approach to the analysis of mode confusion. In *Proc. of the 17th Digital Avionics Systems Conf.*, Bellevue, Washington, USA, 1998.

J.J. Cañas, A. Antolí, and J.F. Quesada. The role of working memory on measuring mental models of physical systems. *Psicológica*, 22:25–42, 2001.

A.R. Cassandra, L.P. Kaelbling, and J.A. Kurien. Acting under Uncertainty: Discrete Bayesian Models for Mobile Robot Navigation. In *Proc. of IEEE/RSJ Int'l. Conf. on Intelligent Robots and Systems (IROS)*, pages 963 – 972, Osaka, Japan, 1996. IEEE.

H. Choset and K. Nagatani. Topological simultaneous localization and mapping (SLAM): toward exact localization without explicit localization. *IEEE Transactions on Robotics and Automation*, 17(2):125 – 136, April 2001.

I.J. Cox. Blanche—An Experiment in Guidance and Navigation of an Autonomous Robot Vehicle. *IEEE Transaction on Robotics and Automation*, 7 (2):193 – 204, April 1991.

J. Crisman and M. Cleary. Progress on the deictically controlled wheelchair. In Mittal et al. (1998), pages 137 – 149.

J. Crow, D. Javaux, and J. Rushby. Models and mechanized methods that integrate human factors into automation design. In K. Abbott, J.-J. Speyer, and G. Boy, editors, *Proc. of the Int. Conference on Human-Computer Interaction in Aeronautics: HCI-Aero 2000*, Toulouse, France, September 2000.

R. Czommer. *Leistungsfähigkeit fahrzeugautonomer Ortungsverfahren auf der Basis von Map-Matching-Techniken*. PhD thesis, Fakultät für Bauingenieur- und Vermessungswesen der Universität Stuttgart, 2001.

A. Degani and M. Heymann. Pilot-autopilot interaction: A formal perspective. In K. Abbott, J.-J. Speyer, and G. Boy, editors, *Proc. of the Int.'l Conf. on Human-Computer Interaction in Aeronautics: HCI–Aero 2000*, pages 157–168, Toulouse, France, September 2000.

A. Degani, M. Shafto, and A. Kirlik. Modes in human-machine systems: Constructs, representation and classification. *Int'l Journal of Aviation Psychology*, 9(2):125–138, 1999.

G. Doherty. *A Pragmatic Approach to the Formal Specification of Interactive Systems*. PhD thesis, University of York, Dept. of Computer Science, October 1998.

D.H. Douglas and T.K. Peucker. Algorithms for the Reduction of the Number of Points Required to Represent a Digitized Line or Its Caricature. *Canadian Cartographer*, 10(2):112 – 122, December 1973.

T. Duckett. *Concurrent Map Building and Self-Localisation for Mobile Robot Navigation*. PhD thesis, Department of Computer Science, University of Manchester, 2000.

Th. Edlinger and G. Weiß. Exploration, navigation and self-localization in an autonomous mobile robot. In R. Dillmann, U. Rembold, and T. Lüth, editors, *Autonome Mobile Systeme*, Informatik aktuell, pages 142–151, Berlin, Heidelberg, New York, 1995. Springer.

A. Elfes. Occupancy grids: A stochastic spatial representation for active robot perception. In S. S. Iyengar and A. Elfes, editors, *Autonomous Mobile Robots*, volume 1, pages 60–70, Los Alamitos, California, 1991. IEEE Computer Society Press.

Encyclopædia Britannica. Online version at http://www.britannica.com, 2002.

S. P. Engelson and D. V. McDermott. Error correction in mobile robot map learning. In *Proceedings of the IEEE Int.'l Conf. on Robotics and Automation*, pages 2555–2560, Nice, France, May 1992. IEEE.

C. Eschenbach, C. Habel, L. Kulik, and A. Leßmöllmann. *Shape Nouns and Shape Concepts: A Geometry for 'Corner'*, volume 1404 of *Lecture Notes in Artificial Intelligence*, pages 177–201. Springer, Berlin, Heidelberg, New York, 1998.

U. Forssell, P. Hall, S. Ahlqvist, and F. Gustafsson. Map-Aided Positioning System. In *Proc. of the Int.'l Federation of Automotive Engineering Societies Congress (FISITA '02)*, Helsinki, Finland, 2002.

D. Fox, W. Burgard, F. Dellaert, and S. Thrun. Monte Carlo localization: Efficient position estimation for mobile robots. In *Proc. of the National Conference on Artificial Intelligence*, 1999.

M.O. Franz and H.A. Mallot. Biomimetic robot navigation. *Robotics and Autonomous Systems, Elsevier*, 30:133 – 153, 2000.

C. R. Gallistel. *The organization of learning*. MIT Press, Cambridge, Massachusetts, 1990.

J.C. García, J. Ureña, M. Mazo, M. Marrón, and S. Palazuelos. Sillas de ruedas autónomas: buses serie e interacción con el entorno. In *Proceedings of IBERDISCAP 2001*, 2001.

T. Gomi and A. Griffith. Developing intelligent wheelchairs for the handicapped. In Mittal et al. (1998), pages 150 – 178.

F. Gustafsson, F. Gunnarsson, N. Bergman, U. Forssell, J. Jansson, R. Karlsson, and P.-J. Nordlund. Particle Filters for Positioning, Navigation and Tracking. *IEEE Transactions on Signal Processing, Special issue on Monte Carlo methods for statistical signal processing*, 2002.

R. Gutierrez-Osuna and R.C. Luo. Lola — Probabilistic Navigation for Topological Maps. *AI magazine*, 17(1), 1995.

J.-S. Gutmann and B. Nebel. Navigation mobiler Roboter mit Laserscans. In P. Levi, Th. Bräunl, and N. Oswald, editors, *Autonome Mobile Systeme*, Informatik aktuell, pages 36–47, Berlin, Heidelberg New York, 1997. Springer.

J.-S. Gutmann, T. Weigel, and B. Nebel. A fast, accurate, and robust method for self-localization in polygonial environments using laser-range-finders. *Advanced Robotics*, 14(8):651 – 668, 2001.

J. Hertzberg and F. Kirchner. Landmark-Based Autonomous Navigation in Sewerage Pipes. In *Proceedings of the First Euromicro Workshop on Advanced Mobile Robots (EUROBOT '96)*, pages 68 – 73. IEEE Press, 1996.

C.A.R. Hoare. *Communicating Sequential Processes*. Prentice-Hall, Englewood Cliffs, New Jersey. USA, 1985.

R. Hourizi and P. Johnson. Beyond mode error: Supporting strategic knowledge structures to enhance cockpit safety. In *Proc. of IHM-HCI 2001*, Lille, France, 2001a. Springer.

R. Hourizi and P. Johnson. Unmasking mode errors: A new application of task knowledge principles to the knowledge gaps in cockpit design. In *Proc. of INTERACT 2001 – The 8th IFIP Conf. on Human Computer Interaction*, Tokyo, Japan, 2001b.

H. Hoyer, U. Borgolte, and A. Jochheim. The omni-wheelchair — state of the art. In *Proc. of the 14th Annual Int. Conf. on Technology and Persons with Disabilities*, Los Angeles, USA, 1999. Online at http://www.dinf.org/csun_99/session0274.html.

E. Hunt and D. Waller. Orientation and wayfinding: A review. ONR Tech. Rep. N00014-96-0380, Office of Naval Research, USA, 1999.

ICRA'02. *Proc. of the 2002 IEEE Int.'l Conf. on Robotics & Automation (ICRA'02)*, Washington DC., USA, May 2002. IEEE, Omnipress, ISBN 0-7803-7273-5.

D. Javaux. Explaining Sarter & Woods' classical results. The cognitive complexity of pilot-autopilot interaction on the Boeing 737-EFIS. In *Proc. of HESSD '98*, pages 62–77, 1998.

D. Javaux. A method for predicting errors when interacting with finite state systems. How implicit learning shapes the user's knowledge of a system. *Reliability Engineering & System Safety*, 72(2):147–165, Feb 2002.

S. Kanowski. Vierter Bericht zur Lage der älteren Generation in der Bundesrepublik Deutschland: Risiken, Lebensqualität und Versorgung Hochaltriger unter besonderer Berücksichtigung demenzieller Erkrankungen. Berlin, January 2002. Deutsches Zentrum für Altersfragen (DZA).

G. Kantor and S. Singh. Preliminary results in range-only localization and mapping. In ICRA'02 ICRA'02, pages 1818 – 1823.

N.I. Katevas, N.M. Sgouros, S.G. Tzafestas, G. Papakonstantinou, P. Beattie, J.M. Bishop, P. Tsanakas, and D. Koutsouris. The autonomous mobile robot SENARIO: A sensor-aided intelligent navigation system for powered wheelchairs. *IEEE Robotics & Automation Magazine*, 4(4):60 – 70, 1997.

S. Kim and J.-H. Kim. Adaptive fuzzy-network-based C-measure map-matching algorithm for car navigation system. *IEEE Transactions on Industrial Electronics*, 48(2):432 – 441, April 2001.

J. Kollmann, A. Lankenau, A. Bühlmeier, B. Krieg-Brückner, and T. Röfer. Navigation of a kinematically restricted wheelchair by the parti-game algorithm. In *Spatial Reasoning in Mobile Robots and Animals*, pages 35 – 44, Manchester University, 1997. AISB-97 Workshop.

J. Kollmann and T. Röfer. Echtzeitkartenaufbau mit einem 180°-Laser-Entfernungssensor. In R. Dillmann, H. Wörn, and M. von Ehr, editors, *Autonome Mobile Systeme 2000*, Informatik aktuell, pages 121–128. Springer, 2000.

B. Krieg-Brückner, T. Röfer, H.-O. Carmesin, and R. Müller. A taxonomy of spatial knowledge and its application to the Bremen Autonomous Wheelchair. In Ch. Freksa, Ch. Habel, and K. F. Wender, editors, *Spatial Cognition*, volume 1404 of *Lecture Notes in Artificial Intelligence*, pages 373–397, Berlin, Heidelberg, New York, 1998. Springer.

B. Kuipers. The spatial semantic hierarchy. *Artificial Intelligence, Elsevier*, 119: 191 – 233, 2000.

B. Kuipers and P. Beeson. Bootstrap learning for place recognition. In *Proc. of the Eighteenth National Conference on Artificial Intelligence (AAAI-02)*, 2002.

B. Kuipers, R. Froom, Y. W. Lee, and D. Pierce. The semantic hierarchy in robot learning. In J. Connell and S. Mahadevan, editors, *Robot Learning*, pages 141–170. Kluwer Academic Publishers, 1993.

B. J. Kuipers and Y.-T. Byun. A robot exploration and mapping strategy based on a semantic hierarchy of spatial representations. *Journal of Robotics and Autonomous Systems*, 8:47–63, 1991.

B.J. Kuipers. *Representing Knowledge of Large-Scale Space*. PhD thesis, Dept. of Mathematics, Massachussets Insititue of Technology, Cambridge, USA, July 1977.

P. Lamb and S. Thiébaux. Avoiding explicit map-matching in vehicle location. In *Proc. of the 6th ITS World Congress (ITS '99)*, Toronto, Canada, November 1999.

A. Lankenau. Avoiding mode confusion in service-robots. In M. Mokhtari, editor, *Integration of Assistive Technology in the Information Age, Proc. of the 7th Int. Conf. on Rehabilitation Robotics*, pages 162 – 167, Evry, France, April 2001. IOS Press.

A. Lankenau and O. Meyer. Der autonome Rollstuhl als sicheres eingebettetes System. Master's thesis, Fachbereich Mathematik und Informatik, Universität Bremen, 1997.

A. Lankenau and O. Meyer. Formal methods in robotics: Fault tree based verification. In *Proc. of Quality Week Europe*, Brussels, Belgium, 1999.

A. Lankenau, O. Meyer, and B. Krieg-Brückner. Safety in robotics: The Bremen Autonomous Wheelchair. In *Proceedings of AMC'98, 5th Int. Workshop on Advanced Motion Control*, pages 524 – 529, Coimbra, Portugal, 1998.

A. Lankenau and T. Röfer. The Role of Shared Control in Service Robots – The Bremen Autonomous Wheelchair as an Example. In Röfer et al. (2000), pages 27 – 31.

A. Lankenau and T. Röfer. Rollstuhl "Rolland" unterstützt ältere und behinderte Menschen. *FIfF-Kommunikation "Informationstechnik und Behinderung"*, (2): 48 – 50, 2000a. Forum InformatikerInnen für Frieden und gesellschaftliche Verantwortung (FIfF) e.V.

A. Lankenau and T. Röfer. Smart Wheelchairs - State of the Art in an Emerging Market. *Künstliche Intelligenz (Heftschwerpunkt Autonome Mobile Systeme)*, (4):37 – 39, 2000b. Gesellschaft für Informatik e.V., arenDTaP Verlag.

A. Lankenau and T. Röfer. The Bremen Autonomous Wheelchair – a versatile and safe mobility assistant. *IEEE Robotics and Automation Magazine, "Reinventing the Wheelchair"*, 7(1):29 – 37, March 2001a.

A. Lankenau and T. Röfer. Selbstlokalisation in Routengraphen. In P. Levi and M. Schanz, editors, *Autonome Mobile Systeme 2001*, Informatik aktuell, pages 157 – 163. Springer, 2001b.

A. Lankenau and T. Röfer. Mobile robot self-localization in large-scale environments. In *Proc. of the Int. Conf. on Robotics and Automation ICRA 2002*, pages 1359 – 1364, Washington, D.C., USA, 2002. IEEE, Omnipress.

A. Lankenau, T. Röfer, and B. Krieg-Brückner. Self-localization in large-scale environments for the bremen autonomous wheelchair. In C. Freksa, W. Brauer, C. Habel, and K.F. Wender, editors, *Spatial Cognition III*, volume 2685 of *Lecture Notes in Artificial Intelligence*, Berlin, Heidelberg, New York, 2002. Springer.

J. C. Laprie, editor. *Dependability: Basic Concepts and Terminology*. Springer, Berlin, Heidelberg, New York, 1992.

J. Leonard and H. F. Durrant-Whyte. Mobile robot localization by tracking geometric beacons. *IEEE Trans. on Robotics and Automation*, 7(3):376–382, June 1991.

N.G. Leveson, L.D. Pinnel, S.D. Sandys, S. Koga, and J.D. Reese. Analyzing software specifications for mode confusion potential. In *Workshop on Human Error and System Development*, Glasgow, UK, 1997.

T.S. Levitt and D.T. Lawton. Qualitative navigation for mobile robots. *Artificial Intelligence*, 44:305 – 360, 1990.

F. Lu and E. Milios. Globally consistent range scan alignment for environment mapping. *Autonomous Robots*, 4:333–349, 1997.

G. Lüttgen and V. Carreño. Analyzing mode confusion via model checking. In D. Dams, R. Gerth, S. Leue, and M. Massink, editors, *SPIN' 99*, volume 1680 of *LNCS*, pages 120–135, Berlin Heidelberg, 1999. Springer.

S. Marsland, U. Nehmzow, and T. Duckett. Learning to select distinctive landmarks for mobile robot navigation. *Robotics and Autonomous Systems, Elsevier*, 37:241 – 260, 2001.

C. Martens, N. Ruchel, O. Lang, O. Ivlev, and A. Gräser. A FRIEND for assisting handicapped people. *IEEE Robotics and Automation Magazine, "Reinventing the Wheelchair"*, 7(1):57 – 65, March 2001.

Merriam-Webster Collegiate Dictionary. Online version at http://www.m-w.com, 2002.

D. Miller. Assistive robotics: An overview. In Mittal et al. (1998), pages 126 – 136.

S.P. Miller and J.N. Potts. Detecting mode confusion through formal modeling and analysis. Technical Report NASA/CR-1999-208971, NASA Langley Research Center, January 1999. Online at http://shemesh.larc.nasa.gov/fm/fm-pubs-larc.html.

V.O. Mittal, H.A. Yanco, J. Aronis, and R. Simpson, editors. *Assistive Technology and AI – Applications in Robotics, User Interfaces, and Natural Language Processing*. Number 1458 in LNAI. Springer, 1998.

R. Müller, T. Röfer, A. Lankenau, A. Musto, K. Stein, and A. Eisenkolb. Coarse qualitative descriptions in robot navigation. In C. Freksa, W. Brauer, C. Habel, and K.F. Wender, editors, *Spatial Cognition II*, volume 1849 of *Lecture Notes in Artificial Intelligence*, pages 265 – 276, Berlin, Heidelberg, New York, 2000. Springer.

A. Mojaev and A. Zell. Online-Positionskorrektur für mobile Roboter durch Korrelation lokaler Gitterkarten. In H. Wörn, R. Dillmann, and D. Henrich, editors, *Autonome Mobile Systeme*, Informatik aktuell, pages 93–99, Berlin, Heidelberg, New York, 1998. Springer.

R. Moratz, B. Nebel, and C. Freksa. Qualitative Spatial Reasoning about Relative Position — The Tradeoff between Strong Formal Properties and Successful Reasoning about Route Graphs. In C. Freksa, W. Brauer, C. Habel, and K.F. Wender, editors, *Spatial Cognition III*, volume 2685 of *Lecture Notes in Artificial Intelligence*, Berlin, Heidelberg, New York, 2002. Springer.

A. Musto. *Qualitative Repräsentation von Bewegungsverläufen*. PhD thesis, Institut für Informatik, Technische Universität München, December 2000.

A. Musto, K. Stein, A. Eisenkolb, and T. Röfer. Qualitative and quantitative representations of locomotion and their application in robot navigation. In *Proc. of the 16th International Joint Conference on Artificial Intelligence (IJCAI-99)*, pages 1067–1073, San Francisco, CA, 1999. Morgan Kaufman Publishers, Inc.

D.A. Norman. Some observations on mental models. In D. Gentner and A.L. Stevens, editors, *Mental Models*. Lawrence Erlbaum Associates Inc., Hillsdale, NJ, USA, 1983.

I. Nourbakhsh, R. Powers, and S. Birchfield. Dervish: An office-navigating robot. *AI Magazine*, 16:53–60, 1995.

C. Nowakowski, Y. Utsui, and P. Green. Navigation System Destination Entry: The Effects of Driver Workload and Input Devices, and Implications for SAE Recommended Practice. Technical report, University of Michigan, Transportation Research Institute (UMTRI), Ann Arbor, USA, May 2000.

M. Nuttin, E. Demeester, D. Vanhooydonck, and H. Van Brussel. Obstacle Avoidance and Shared Autonomy for Electric Wheel Chairs: An Evaluation of Preliminary Experiments. In *Robotik 2002*, number 1679 in VDI-Berichte, pages

509 – 513, Ludwigsburg, Germany, jun 2002. VDI/VDE-Gesellschaft Mess-
und Automatisierungstechnik, VDI-Verlag GmbH, Düsseldorf.

P. Odor. *The CALL Centre Smart Wheelchair.* CALL Centre, 1995.

E. Palmer. "Oops, it didn't arm." – A case study of two automation surprises.
In *Proc. of the 8th Int'l Symp. on Aviation Psychology*, Ohio State University,
Columbus, USA, 1995.

D.L. Parnas and J. Madey. Functional documents for computer systems. *Science
of Computer Programming*, 25(1):41–61, October 1995.

G. Pires, R. Araújo, U. Nunes, and A.T. de Almeida. Robchair – a powered wheelchair using a behaviour-based navigation. In *Proc. of AMC'98, 5th Int. Workshop on Advanced Motion Control*, pages 536 – 541, Coimbra, Portugal, 1998.

E. Prassler, J. Scholz, and P. Fiorini. A Robotic Wheelchair for Crowded Public
Environments. *IEEE Robotics and Automation Magazine, "Reinventing the
Wheelchair"*, 7(1):38 – 45, March 2001.

E. Remolina and B. Kuipers. Towards a general theory of topological maps.
Technical report, Computer Science Department, University of Texas at Austin,
USA, 2002.

B. Richardson and P. Green. Trends in north american intelligent transportation systems: A year 2000 appraisal. Technical report, University of Michigan,
Transportation Research Institute (UMTRI), Ann Arbor, USA, April 2000.

M. Rodriguez, M. Zimmermann, M. Katahira, M. de Villepin, B. Ingram, and
N. Leveson. Identifying mode confusion potential in software design. In *Proc.
of the Int. Conference on Digital Aviation Systems*, Philadelphia, PA, USA, October 2000.

T. Röfer. Navigation mit eindimensionalen 360° Bildern. In R. Dillmann, U. Rembold, and T. Lüth, editors, *Autonome Mobile Systeme*, Informatik aktuell, pages
193–202, Berlin, Heidelberg, New York, 1995. Springer.

T. Röfer. Controlling a wheelchair with image-based homing. In *Spatial Reasoning in Mobile Robots and Animals*, pages 66–75, Manchester University, 1997a.
AISB-97 Workshop.

T. Röfer. Routemark-based navigation of a wheelchair. In *Proc. Third ECPD Int.'l
Conf. on Advanced Robotics, Intelligent Automation and Active Systems*, pages
333–338, Bremen, 1997b.

T. Röfer. *Panoramic Image Processing and Route Navigation. Ph. D. thesis*, volume 7 of *BISS Monographs*. Shaker, 1998a.

T. Röfer. Routenbeschreibung durch Odometrie-Scans. In H. Wörn, R. Dillmann, and D. Henrich, editors, *Autonome Mobile Systeme 1998*, Informatik aktuell, pages 122–129, Berlin, Heidelberg, New York, 1998b. Springer.

T. Röfer. Strategies for using a simulation in the development of the Bremen Autonomous Wheelchair. In R. Zobel and D. Moeller, editors, *Simulation-Past, Present and Future*, pages 460–464. Society for Computer Simulation International, 1998c.

T. Röfer. Route navigation and panoramic image processing. In *Ausgezeichnete Informatikdissertationen 1998*, pages 132–141, Leipzig, 1999a. B. G. Teubner Stuttgart.

T. Röfer. Route navigation using motion analysis. In *Proc. Conf. on Spatial Information Theory '99*, volume 1661 of *Lecture Notes in Artificial Intelligence*, pages 21–36, Berlin, Heidelberg, New York, 1999b. Springer.

T. Röfer. Building consistent laser scan maps. In *Proc. of the 4th European Workshop on Advanced Mobile Robots (Eurobot 2001)*, volume 86 of *Lund University Cognitive Studies*, pages 83 – 90, 2001a.

T. Röfer. Konsistente Karten aus Laser Scans. In *Autonome Mobile Systeme 2001*, Informatik aktuell, pages 171–177. Springer, 2001b.

T. Röfer. Using Histogram Correlation to Create Consistent Laser Scan Maps. In *Proc. of the IEEE Int'l Conf. on Robotics Systems (IROS-2002)*, pages 625 – 630, EPFL, Lausanne, Switzerland, September 2002. IEEE.

T. Röfer and A. Lankenau. Rolland's web pages including media echo. URL http://www.informatik.uni-bremen.de/rolland.

T. Röfer and A. Lankenau. Architecture and applications of the Bremen Autonomous Wheelchair. In P. P. Wang, editor, *Proc. of the Fourth Joint Conference on Information Systems*, volume 1, pages 365 – 368. Association for Intelligent Machinery, 1998.

T. Röfer and A. Lankenau. Ein Fahrassistent für ältere und behinderte Menschen. In *Autonome Mobile Systeme 1999*, Informatik aktuell, pages 334 – 343, Berlin, Heidelberg, New York, 1999a. Springer.

T. Röfer and A. Lankenau. Ensuring safe obstacle avoidance in a shared-control system. In J. M. Fuertes, editor, *Proc. of the 7th Int. Conf. on Emergent Technologies and Factory Automation*, pages 1405 – 1414, 1999b.

T. Röfer and A. Lankenau. Architecture and applications of the Bremen Autonomous Wheelchair. *Information Sciences*, 126(1-4):1 – 20, July 2000.

T. Röfer and A. Lankenau. Route-based robot navigation. *Freksa, C. (Hrsg.):* Künstliche Intelligenz (Themenheft Spatial Cognition), pages 29 – 31, 2002. Fachbereich 1 der Gesellschaft für Informatik e.V., arenDTaP Verlag.

T. Röfer, A. Lankenau, and R. Moratz, editors. *Workshop "Service Robotics – Applications and Safety Issues in an Emerging Market"*, Berlin, Germany, 2000. European Conference on Artificial Intelligence (ECAI 2000).

T. Röfer and R. Müller. Navigation and routemark detection of the Bremen Autonomous Wheelchair. In T. Lüth, R. Dillmann, P. Dario, and H. Wörn, editors, *Distributed Autonomous Robotics Systems*, pages 183–192, Berlin, Heidelberg, New York, 1998. Springer.

S. Rogers, C.-N. Flechter, and P. Langley. An adaptive interactive agent for route advice. In O. Etzioni, J.P. Müller, and J.M. Bradshaw, editors, *Proceedings of the Third Int.'l Conf. on Autonomous Agents (Agents'99)*, pages 198 – 205, Seattle, WA, USA, 1999. ACM Press. URL http://citeseer.nj.nec.com/rogers95adaptive.html.

A. W. Roscoe. *The Theory and Practice of Concurrency*. Prentice-Hall, 1997. ISBN 0-13-674409-5.

J. Rushby. Analyzing cockpit interfaces using formal methods. In H. Bowman, editor, *Proc. of FM-Elsewhere*, volume 43 of *Electronic Notes in Theoretical Computer Science*, Pisa, Italy, October 2000. Elsevier.

J. Rushby. Modeling the human in human factors. In *Proc. of SAFECOMP 2001*, volume 2187 of *LNCS*, pages 86–91. Springer, 2001.

J. Rushby. Using model checking to help discover mode confusions and other automation surprises. *Reliability Engineering & System Safety*, 75(2):167 – 177, Feb 2002.

J. Rushby, J. Crow, and E. Palmer. An automated method to detect potential mode confusions. In *Proc. of the 18th AIAA/IEEE Digital Avionics Systems Conf.*, St. Louis, Montana, USA, 1999.

J.S. Russell and P. Norvig. *Artificial Intelligence: A Modern Approach.* Prentice-Hall, New Jersey, USA, 1995.

M. Sage and C.W. Johnson. Formally verified, rapid prototyping for air traffic control. *Reliability Engineering & System Safety*, 72(2):121–132, Feb 2002.

N. Sarter and D. Woods. How in the world did we ever get into that mode? Mode error and awareness in supervisory control. *Human Factors*, 37(1):5–19, 1995.

M.A. Sasse. *Eliciting and Describing Users' Models of Computer Systems.* PhD thesis, School of Computer Science, The University of Birmingham, Apr 1997. Online at http://www.cs.ucl.ac.uk/staff/A.Sasse/thesis/LINK_ME.html.

K. Schilling and H. Roth. Convoy driving and obstacle avoidance systems for electrical wheelchairs. In P.P. Wang, editor, *Proc. of the 4th Joint Conf. on Information Systems*, volume 1, pages 353 – 356. Association for Intelligent Machinery, 1998.

A.J. van Schouwen, D.L. Parnas, and J. Madey. Documentation of requirements for computer systems. In *IEEE Int'l. Symp. on Requirements Engineering – RE'93*, pages 198–207, San Diego, California, USA, 4–6 Jan. 1993. IEEE Comp. Soc. Press.

U. Siems, Ch. Herwig, and T. Röfer. Simrobot - Ein Programm zur Simulation sensorbestückter Agenten in einer dreidimensionalen Umwelt. In B. Krieg-Brückner, G. Roth, and H. Schwegler, editors, *ZKW-Bericht*, number 1/94 in ISSN 0947-0204, Universität Bremen, 1994. Zentrum für Kognitionswissenschaften. Online at http://www.tzi.org/~simrobot.

R. Simmons and S. Koenig. Probabilistic robot navigation in partially observable environments. In *Proc. of the Int. Joint Conf. on Artificial Intelligence, IJCAI-95*, pages 1080–1087, 1995.

R.C. Simpson, S.P. Levine, D.A. Bell, L.A. Jaros, Y. Koren, and J. Borenstein. Navchair: An assistive wheelchair navigation system with automatic adaptation. In Mittal et al. (1998), pages 235 – 255.

N. Storey. *Safety-Critical Computer Systems.* Addison-Wesley, 1996.

H. Thimbleby. *User Interface Design.* ACM Press, New York (USA), 1990.

S. Thrun. Learning maps for indoor mobile robot navigation. *Artificial Intelligence*, 99:21 – 71, 1998.

S. Thrun, W. Burgard, and D. Fox. A Real-Time Algorithm for Mobile Robot Mapping With Applications to Multi-Robot and 3D Mapping. In *Proc. of the IEEE Int. Conf. on Robotics & Automation*, pages 321 – 328, 2000a.

S. Thrun, D. Fox, W. Burgard, and F. Dellaert. Robust Monte Carlo localization for mobile robots. *Artificial Intelligence*, 101:99 – 141, 2000b.

N. Tomatis. *Hybrid, Metric-Topological, Mobile Robot Navigation*. PhD thesis, École Polytechnique Fédérale de Lausanne, 2001.

N. Tomatis, I. Nourbakhsh, and R. Siegwart. Combining Topological and Metric: A Natural Integration for Simultaneous Localization and Map Building. In *Proc. of the 4th European Workshop on Advanced Mobile Robots (Eurobot 2001)*, volume 86 of *Lund University Cognitive Studies*, Sweden, 2001a.

N. Tomatis, I. Nourbakhsh, and R. Siegwart. Simultaneous localization and map building: A global topological model with local metric maps. In *Proceedings of the IEEE/RSJ Int.'l Conf. on Intelligent Robots and Systems (IROS 2001)*, Maui, Hawaii, October 2001b.

N. Tomatis, I. Nourbakhsh, and R. Siegwart. Hybrid Simultaneous Localization and Map Building: Closing the Loop with Multi-Hypotheses Tracking. In ICRA'02 ICRA'02, pages 2749 – 2754.

O. Trullier, S. I. Wiener, A. Bertholz, and J.-A. Meyer. Biologically based artificial navigation systems: Review and prospects. In *Progress in Neurobiology*, volume 51, pages 483–544. Pergamon, 1997.

I. Ulrich and I. Nourbakhsh. Appearance-Based Place Recognition for Topological Localization. In *Proc. of the IEEE Int.'l Conf. on Robotics and Automation*, San Francisco, CA, USA, April 2000. IEEE. Best Vision Paper Award.

UN/ECE(1999). United Nations Economic Commission for Europe (un/ece): World robotics 1999 – statistics, market analysis, forecasts, case studies and profitability of robot investment. ISBN 92-1-101007-1, 1999. Short version as UN/ECE press release ECE/STAT/99/2 online at http://www.unece.org/press/99stat2e.htm.

UN/ECE(2001). United Nations Economic Commission for Europe (un/ece): World robotics 2001 – statistics, market analysis, forecasts, case studies and profitability of robot investment. ISBN 92-1-101043-8, 2001. Short version as UN/ECE press release ECE/STAT/01/10 online at http://www.unece.org/press/01stat10.htm.

J. Ureña, J.J. García, E. Bueno, M. Mazo, Á. Hernández, J.C. García, and V. Díaz. New sonar configuration for a powered wheelchair. In J. M. Fuertes, editor, *Proc. of the 7th Int. Conf. on Emergent Technologies and Factory Automation*, pages 113 – 119, 1999.

S.S. Vakil and R.J. Hansman, Jr. Approaches to mitigating complexity-driven issues in commercial autoflight systems. *Reliability Engineering & System Safety*, 72(2):133–145, Feb 2002.

P. van Oosterom. *Reactive Data Structures for Geographic Information Systems*. PhD thesis, Rijksuniversiteit te Leiden, Leiden, The Netherlands, 1991.

VRML97. The virtual reality modeling language, international standard iso/iec 14772-1:1997, 1997.

R. Wehner. Middle-scale navigation: The insect case. *Journal of Experimental Biology*, 199, 1996.

R. Wehner. Large-Scale Navigation: The Insect Case. In *Proc. Conf. on Spatial Information Theory '99*, volume 1661 of *Lecture Notes in Artificial Intelligence*, pages 1–20, Berlin, Heidelberg, New York, 1999. Springer.

S. Werner, B. Krieg-Brückner, and Th. Herrmann. *Modelling Navigational Knowledge by Route Graphs*, volume 1849 of *Lecture Notes in Artificial Intelligence*, pages 295–316. Springer, Berlin, Heidelberg, New York, 2000.

S. Werner, B. Krieg-Brückner, H. A. Mallot, K. Schweizer, and Ch. Freksa. Spatial cognition: The role of landmark, route, and survey knowledge in human and robot navigation. In *Informatik '97 – Informatik als Innovationsmotor*, Informatik aktuell, Berlin, Heidelberg, New York, 1997. Springer.

S.D. Whitehead and D.H. Ballard. Learning to Perceive and Act by Trial and Error. *Machine Learning*, 7:45–83, 1991.

P. Wright, B. Fields, and M. Harrison. Deriving human-error tolerance requirements from tasks. In *Proc. of the 1st Int. Conf. on Requirements Engineering*, pages 135–142, Colorado, USA, 1994. IEEE.

H. Yanco. Wheelesley: A robotic wheelchair system: Indoor navigation and user interface. In Mittal et al. (1998), pages 256 – 268.

Y. Zhao. *Vehicle Location and Navigation Systems*. Artech House Intelligent Transportation Systems Library. Artech House, Inc., Norwood, MA, USA, 1997.

M. Zimmermann, M. Rodriguez, B. Ingram, M. Katahira, M. de Villepin, and N. Leveson. Making formal methods practical. In *Proc. of the Int. Conference on Digital Aviation Systems*, Philadelphia, PA, USA, October 2000.

D. van Zwynsvoorde, T. Simeon, and R. Alami. Incremental topological modeling using local Voronoï-like graphs. In *Proc. of IEEE/RSJ Int. Conf. on Intelligent Robots and System (IROS 2000)*, volume 2, pages 897 – 902, Takamatsu, Japan, October 2000.

D. van Zwynsvoorde, T. Simeon, and R. Alami. Building topological models for navigation in large scale environments. In *Proc. of IEEE Int. Conf. on Robotics and Automation ICRA 2001*, pages 4256 – 4261, Seoul, Korea, May 2001.

Monographs of the Bremen Institute of Safe Systems

1   Buth, Bettina / Berghammer, Rudolf / Peleska, Jan (eds.): *Tools for System Development and Verification*. Workshop, Proceedings, Bremen, Germany, July 1996. ISBN 3-8265-3806-4. Aachen: Shaker Verlag, 1998.

2   Mossakowski, Till: *Representations, Hierarchies and Graphs of Institutions*. Dissertation, Universität Bremen, 1996. Revised version: ISBN 3-89722-831-9. Berlin: Logos Verlag, 2001.

3   Cerioli, Maura / Gogolla, Martin / Kirchner, Hélène / Krieg-Brückner, Bernd / Qian, Zhenyu / Wolf, Markus (eds.): *Algebraic System Specification and Development: Survey and Annotated Bibliography*. 2nd edition, 1997. ISBN 3-8265-4067-0. Aachen: Shaker Verlag, 1998.

4   Wolff, Burkhart: *A Generic Calculus of Transformation*. Dissertation, Universität Bremen, 1997. Revised version: ISBN 3-8265-3654-1. Aachen: Shaker Verlag, 1999.

5   Kolyang: *HOL-Z, An Integrated Formal Support Environment for Z in Isabelle/HOL*. Dissertation, Universität Bremen, 1997. ISBN 3-8265-4068-9. Aachen: Shaker Verlag, 1998.

6   Fröhlich, Michael: *Inkrementelles Graphlayout im Visualisierungssystem daVinci*. Dissertation, Universität Bremen, 1997. ISBN 3-8265-4069-7. Aachen: Shaker Verlag, 1998.

7   Röfer, Thomas: *Panoramic Image Processing and Route Navigation*. Dissertation, Universität Bremen, 1998. ISBN 3-8265-4070-0. Aachen: Shaker Verlag, 1998.

8   Schrönen, Michael: *Methodology for the Development of Microprocessor-Based Safety-Critical Systems*. Dissertation, Universität Bremen, 1998. ISBN 3-8265-3655-X. Aachen: Shaker Verlag, 1998.

9   Krieg-Brückner, Bernd / Peleska, Jan / Olderog, Ernst-Rüdiger / Balzer, Dietrich / Baer, Alexander: *UniForM Workbench, Universelle Entwicklungsumgebung für Formale Methoden; Schlußbericht*. ISBN 3-8265-3656-8. Aachen: Shaker Verlag, 1999.

10 Gärtner, Heino: *Schematransformationen in objektorientierten Informationssystemen*. Dissertation, Universität Bremen, 1999. ISBN 3-8265-6542-8. Aachen: Shaker Verlag, 1999.

11 Huge, Anne-Kathrin: *Ein Ansatz zur Formalisierung objektorientierter Datenbanken auf der Grundlage von ODMG*. Dissertation, Universität Bremen, 1999. ISBN 3-8265-6543-6. Aachen: Shaker Verlag, 2000.

12 Karlsen, Einar: *Tool Integration in a Functional Programming Language*. Dissertation, Universität Bremen, 1998. Revised version. Universität Bremen, 1999.

13 Amthor, Peter: *Structural Decomposition of Hybrid Systems*. Dissertation, Universität Bremen, 2000.

14 Richters, Mark: *A Precise Approach to Validating UML Models and OCL Constraints*. Dissertation, Universität Bremen, 2001. ISBN 3–89722-842-4. Berlin: Logos Verlag, 2002.

15 Buth, Bettina: *Formal and Semi-Formal Methods for the Analysis of Industrial Control Systems*. Habilitationsschrift, Universität Bremen, 2001. Berlin: Logos Verlag, 2002.

16 Meyer, Oliver: *Structural Decomposition of Timed CSP and its Application in Real-Time Testing*. Dissertation, Universität Bremen, 2001. Berlin: Logos Verlag, 2002.

17 Kollmann, Ralf: *Design Recovery Techniques for Object-Oriented Software Systems*. Dissertation, Universität Bremen, 2002. ISBN 3–8325-0141-X. Berlin: Logos Verlag, 2002.

18 Lankenau, Axel: *The Bremen Autonomous Wheelchair "Rolland": Self-Localization and Shared Control*. Dissertation, Universität Bremen, 2002. ISBN 3–8325-0286-6. Berlin: Logos Verlag, 2003.

19 Tej, Haykal: *HOL-CSP: Mechanised Formal Development of Concurrent Processes*. Dissertation, Universität Bremen, 2003. ISBN 3–8325-0287-4. Berlin: Logos Verlag, 2003.